U0270583

清华大学建筑 规划 景观设计教学丛书
课程编号：70000614

印象北京 / 巴塞罗那

——颐和园西南区域及黄石铁山区矿坑景观设计

朱育帆　邬东璠　编著

中国建筑工业出版社

图书在版编目（CIP）数据

印象北京／巴塞罗那——颐和园西南区域及黄石铁山区矿坑景观设计／朱育帆，邬东播编著.—北京：中国建筑工业出版社，2016.10
（清华大学建筑 规划 景观设计教学丛书）
ISBN 978-7-112-19875-7

Ⅰ.①印… Ⅱ.①朱…②邬… Ⅲ.①景观设计－作品集－世界－现代 Ⅳ.①TU983

中国版本图书馆CIP数据核字（2016）第220382号

本书是清华大学建筑学院70周年院庆图书，本书中包含的设计课程场地有颐和园西南区和黄石铁山区矿坑两个选点，前者重点在安置必要生态功能、水利功能与妥善解决居民生活问题。后者主要是工业历史遗迹转型旅游开发。

责任编辑：陈 桦 杨 琪
责任校对：李欣慰 姜小莲

清华大学建筑 规划 景观设计教学丛书
印象北京/巴塞罗那——颐和园西南区域及黄石铁山区矿坑景观设计
朱育帆 邬东播 编著

*

中国建筑工业出版社出版、发行（北京西郊百万庄）
各地新华书店、建筑书店经销
北京京点图文设计有限公司制版
北京缤索印刷有限公司印刷

*

开本：889×1194毫米 1/20 印张：13⅕ 字数：300千字
2016年9月第一版 2016年9月第一次印刷
定价：85.00元
ISBN 978-7-112-19875-7
（29408）
版权所有 翻印必究
如有印装质量问题，可寄本社退换
（邮政编码 100037）

本书编委会

主　　编：朱育帆　邬东璠

编　　委：李　宾　吕　回　李正祥　边思敏

参　　编：朱一君　梁思佳　冯　阳　罗　茜
武　鑫　赵婷婷　孙媛娜　解陈娟
马双枝　李宏丽　慕晓东　童　牧
杨永亮　赵维佳　刘　畅

前　言

2011 年 2 月学院安排我作为访问教授前往巴塞罗那加泰罗尼亚理工大学（UPC）建筑学院（ETSAB）教授中国园林史和景观设计，在时任副院长Kari女士的引荐下，我第一次见到了米盖尔・比达尔(Miquel Vital)教授，工作是协助他讲授一门研究生的景观设计课程。驼背，慢速，和蔼、浓重难懂的加泰罗尼亚式英语这些都是当时对于老先生的第一印象，然而就是这样一位貌似普通的老人，在这里是一定要费些笔墨的，因为没有他的坚持可能就没有清华景观设计课程联合教学国际化的今天了。

米盖尔教授出生在一个典型的加泰罗尼亚知识分子家庭，父亲是加泰知名的艺术家和建筑师，他子承父业，如今他的女儿又再承父业。从米盖尔的日常着装可以看出，他有着极高的艺术修养和品味，帽子、围巾、袜子，配饰等的选择与整体形象完全经过仔细考量和推敲，浓浓的艺术范儿。米盖尔知识渊博，熟知巴塞的历史，对这座城市寸草片瓦皆有深情，思想和举止中渗透着加泰式的傲骨感。 他的身体其实非常强壮，有着惊人的耐力和毅力，几乎所有事情都身体力行，不知疲倦，让晚辈的我们自愧不如。

如今渐为国内所认知的欧洲最为重要的景观奖是以早逝的罗莎・芭芭拉（Rosa Barba）为命名的，米盖尔和罗莎是同事，与其强调大生态的景观教育理念不同，他更执着于景观与文化与艺术之间的关联，在 ETSAB 米盖尔对专业教育所投入的巨大热忱是独一无二的，这一点不由得让人肃然起敬。

在 ETSAB 授课接近尾声的时候碰巧遇到建筑学院党安荣老师和郑晓笛老师带领清华景观规划 studio 的学生许愿、蒙宇婧、高飞、倪小漪、龙漩、贾崇俊、梁大庆、莫珊，杨曦和杨洋等到访巴塞罗那，规划课程是以西班牙 Fayón 小镇景观复兴为题，Fayón 老镇曾是采煤重镇，后因水库建设被淹没一蹶不振，场地集地质、水文、后工业遗存和集体记忆于一身，中西学生在 ETSAB 的 workshop 非常成功，在和米盖尔的交流中他提到了对清华景观教学方式的困惑，他发现双方所关心的知识点好像完全不同，清华方面侧重于宏观的涉及政策策略、旅游规划和经济分析等，而他侧重的是如何通过在场获得感性的和直觉的艺术灵感，米盖尔最终意识到他所带的设计课程与清华的景观设计 studio 对接才是对位的教学合作。自这次 workshop 之后米盖尔将自己的设计课程调整到秋季学期与清华的景观设计课程同步，同年 12 月份他带领西班牙学生组造访清华大学，完成了双方第一次的设计 workshop，并逐渐形成了以同一课题和互访 workshop 为基本标志的联合设计模式。经过一年的筹划 2012 年清华和 UPC 实现了真正意义的互访，9 月 UPC 课程组在清华大学建筑学院举办“印象巴塞罗那”的专题展览，西班牙大使馆参赞和清华大学副校长到场祝贺，11 月清华回访在 UPC 建筑学院举办“印象北京”的专题展览，都获得了很大的成功。米盖尔制定了缜密详尽的在巴塞罗那期间的教学计划，从接机到安排重量级的学者授课，巴塞罗那城市 field trip，城市周边典型项目的 field trip，集中 workshop 工作营，final review ，全体师生家宴直至送机的“套餐式”合作。对于课程米盖尔事必躬亲，一丝不苟，始终如一，学生们都尊称“米爷爷”，在多年连续的合作中米盖尔展现了一贯的沉稳，善良，执着和真诚的品性，赢得了师生的尊重和爱戴。

联合设计课程还得到了清华建筑学院领导层朱文一院长、庄惟敏院长、刘健副院长的大力支持，张利教授、刘畅教授、王贵祥教授、刘健副院长、李树华老师、党安荣老师和胡洁老师等重量级学者亲自授课，景观学系除了我和邬东璠作为课程指导教师之外，庄优波、刘海龙老师和郑晓笛老师都参与了课程评图。在与 UPC 的合作中，清华景观设计教育从理念、意识到技术都得到了长足的提高，收获满满。

不曾想与米盖尔的合作自 2011 年起有着续写至今的缘份，这本册子是关于双方前 3 次合作的集成，主题分别是印象北京 / 巴塞罗那，颐和园西南区域，黄石铁山区后工业景观改造。最后要感谢博士后张安、高杰、郭湧和沈洁对于课程顺利运转的默默贡献，还要特别感谢博士生李宾、吕回、边思敏，硕士生李正祥在书籍组织排版编写中的艰苦付出，清华景观教育规模甚小，高效和众志成城是收获成果的关键。

朱育帆

北京海淀区学清苑一语云舍

Preface

Feb, 2011, I was sent by the university as a visiting professor of ETSAB, Polytechnic University of Catalonia, teaching Chinese garden history and landscape design. Under the recommendation of Mrs. Kari, I worked as an assistant for Professor Miquel Vidal with landscape design course for graduate students. My first impression for Professor Miquel Vidal is more or less like humpback but very kind, slow voice and difficult Catalonian English. However, I have to talk at length about this ordinary looking old man here, because without his continuing efforts, it is impossible to see all the successful joint landscape design education in Tsinghua today.

Professor Miquel Vidal comes from an intelligent Catalonian family , his father is a well-known artist and architect, then Miquel made himself an architect later on while now his daughter works as an architect as well. Miquel has a very good taste for art, because his hats, scarves, socks, and other accessories are carefully chosen for each designed artistic look. Miquel also has a very good knowledge for history of Barcelona, and he is deeply in love with every single thing around the city. There is a distinguished Catalonian elegancy that can be tell by Miquel's words and behaviors. Meanwhile, Miquel is very tough and strong as a professor, with very impressive patience and perseverance always, and almost did everything on his own, tireless and even more energetic than the younger people, which embarrasses us a lot.

As we may all know that the most renowned landscape architecture award in Europe is named after Rosa Barba. Professor Miquel Vidal was once working very closely with Rosa Barba. However, what is different from Rosa Barba's ecological advocacy, Miquel is more obsessed with the connection between landscape and culture & art. Here in ETSAB, the tremendous efforts and enthusiasm that Professor Miquel has put into the landscape design education owns him very high reputation in the profession.

At the end of the ETSAB studio, we happened to meet professor DANG Anrong and professor ZHENG Xiaodi from Tsinghua who was leading a studio field trip to Barcelona, the students group whose name are XU Yuan, MENG Yujing, GAO Fei, NI Xiaoyi, LONG Xuan, JIA Chongjun, LIANG Daqing, MO Shan, YANG Xi and YANG Yang, were working on a planning studio on town landscape reVidalization of Fayón, Spain The old town Fayón was once a mining town, and then flooded down due to the reservoir construction. A collective memory of Fayón's geology, hydrology and industrial heritage deeply resides on site. It was a great success on cross-cultural communication of the ETSAB workshop. But Professor Miquel Vidal also found some confusions about landscape architecture education in Tsinghua , for example, he thought Tinghua focus mainly on large scale landscape issues, such as policy strategy, tourism planning and economic analysis other than paying more attention on intuitively design thinking from site. In the end, Miquel realized that only by collaborating design studio with Tsinghua probably was the only way that would work. Immediately after this workshop, Professor Miquel changed his design studio to the fall in order to coordinate with Tsinghua's studio time schedule. Thus In the following December, Miquel lead his students group to Tsinghua university, and we successfully accomplished THU-UPC joint workshop that year for the first time. Finally we all agree on a jointed design-education model of working on same design issues together and encouraging exchange studio visits. After one-year's plan, Tsinghua and UPC launched the studio visiting program on 2012. On September, the UPC course panel held a special exhibition for "Barcelona impression" in THU Architecture Department. Embassy Counsellor of Spain and Vice-President of Tsinghua University both came for the celebration ceremony. Two months later in November, UPC Architecture School also held a special exhibition of "Beijing impression" during the time of our visit back. These two exhibition are very successful. Professor Miquel Vidal had made a plans packaged for this collaboration work, which includes a very deliberative teaching plan in Barcelona, covering from airport pick up errand to the selections of lecturers , then the Barcelona field trip and trip to specific sites, workshop schedule , final review, and even the final faculty dinner. Miquel did everything on his own as much as he could, the students called him "Grandpa Mi" . His kindness, maturing work style, sincerity and delicate attitude wins him a very good reputation.

The jointed workshop also received strong supports from the dean's panel here at THU architecture department. Dean ZHU Wenyi, ZHUANG Weiming, LIU Jian and Professor LI Shuhua, DANG Anrong and HU Jie assisted with the jointed design studio, professor ZHUANG Youbo, LIU Hailong and ZHENG Xiaodi are invited as design critics for the workshop review. Tsinghua landscape design education gains a lot in design idea, concept and technique from this jointed efforts with UPC.

I never expect that our collaboration work with Miquel Vidal goes such a long way from 2011 till now. And this book is a summary about all these three collaborations, themes by Beijing impression/Barcelona, Southwest summer palace, Tieshan area of Huangshi post industry landscape transformation. Also, I would like to thank my PHD students LI Bing, LYU Hui, Master degree student LI Zhengxiang, I appreciate all their efforts for the publication of this book. THULA is not a very big group, only by working effectively and coherently enable us to achieve all the goals in the end.

Zhu Yufan

at Yiyu Yunshe, Beijing

(Translator: Lyu Hui)

目　录

课程简介 Brief Introduction of Studio

印象北京 / 巴塞罗那展

加泰罗尼亚理工大学与清华大学两校师生通过对两个城市的文化、风俗的仔细走访研究，分别对其城市化进程、景观等做出了深入的分析，希望学生们触及并真正懂得到设计的本质问题，以此为基础来有效加深两个处于不同地域学校间的交流，同时达到拓宽学生们的设计知识面的目的。
Year byear the contact and exchange between professors and students of oriental and occidental universities is more frequent.However this exchange often does not deepen sufficiently enough into the different cultures and civilizations. It is more concentrated on concrete or technological aspects that are essentially neutral.Maybe in the fields of Urbanism and Landscape, this necessity to broaden the knowledge of identities, and the intangibles that generate them, seems more evident.

三校联合 STUDIO

为了探讨参数化设计与场地的问题，形成清华大学分别与加泰罗尼亚理工大学和英国建筑联盟学院形成了联合设计课程。
选点一（2012 年）位于颐和园西南侧，毗邻京郊三山五园，地势低洼水量充沛，历史上有着优良的自然条件。
随着城市的不断扩展，原本肥沃的农田逐渐被城中村和苗圃取代，生态环境被逐步破坏；场地在未来若干年内会被改建成南水北调工程终点的蓄水池，如何安置必要生态功能、水利功能，同时妥善解决居民生活问题，成为设计的要点。
选点二（2013 年）是黄石市工业遗址中巨大的天坑。该场地面临资源枯竭、经济结构改变。转型旅游是这里发展的一个机遇。保留工业历史遗迹是开发旅游的立足点。对于这座城市来说，这是未来的财富。
In order to explore the problem of parametrization design and regional characteristics, THU cooperated and launched joint workshop & studio with UPC and AA respectively.
The site (2012) locates by the north-west side of the Summer Palace, where the amount of water was once very rich and many traditional royal gardens were built since 17th century.
However, due to Beijing's urban sprawl, the great ecological circumstance in history was replaced by illegal residence and nursery gardens. Also the government planned to drive off the residents in the site and build the terminal reservoir of the South-to-North Water Diversion Project. The ecological damage, the dwelling of the inhabitants and the Hydraulic engineering became the main issues of this site.
Site(2013) huge sinkhole is one of the most important Huangshi industrial remains. The site faces the resource depletion the economic structural change.Transformation for tourism is an opportunity of the city.

清华大学建筑学院

School of Architecture, Tsinghua University

朱育帆 教授
景观系副系主任
博士生导师

邬东璠 副教授
硕士生导师

张安 博士后

郭湧 博士后

加泰罗尼亚理工大学建筑学院

School of Architecture, UPC

Miquel Vidal Pla
教授
博士生导师

Pamela Duran
博士后

英国 AA 建筑学院

Architectural Association School of Architecture

Eva Castro 教授
Plasma 事务所创始人

Nicola Saladino
reMIX 事务所创始人

Federico Ruberto
reMIX 事务所创始人

印象北京 / 巴塞罗那 展览照片

2012.9~2012.11

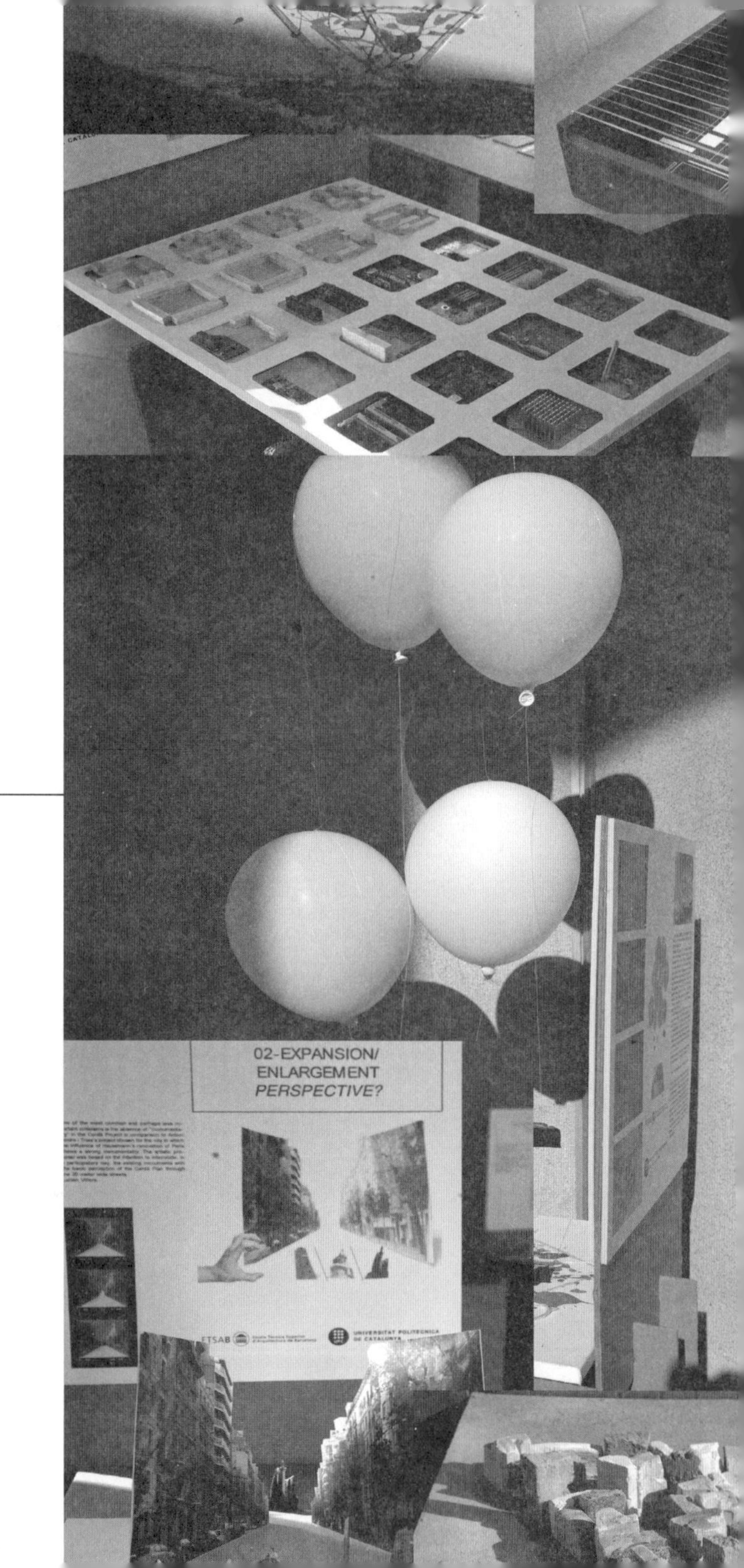

01 开幕仪式 Opening Ceremony

印象北京 / 巴塞罗那展览开幕典礼于 2012 年 9 月在清华大学建筑学院举行，两校师生参与了本学期的联合设计工作营。

开幕典礼各校老师致辞

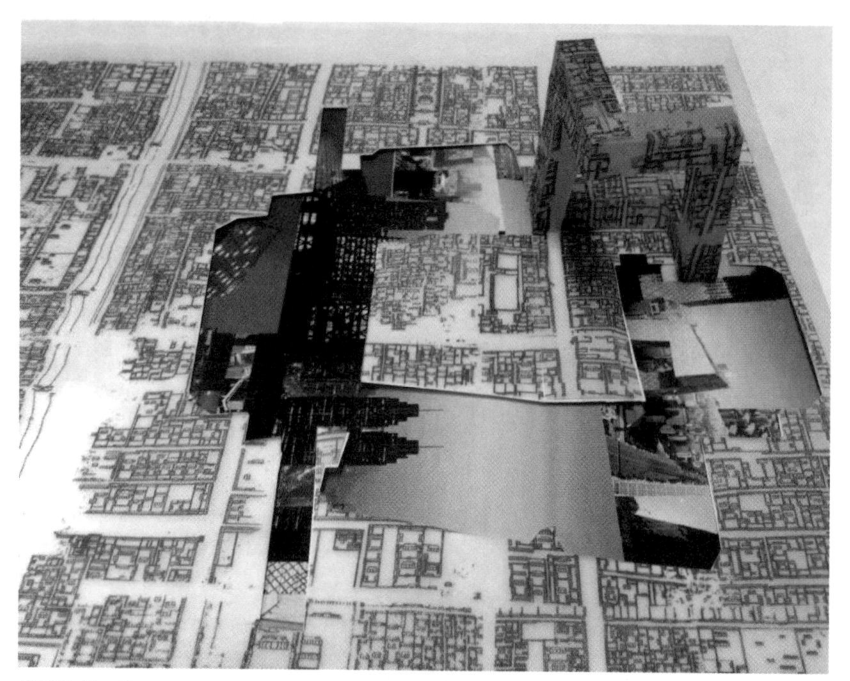

模型表达：Zhao Tingting/Sun Yuanna/Wu Xin

模型表达：Liu Chang/Zhao Weijia/Liang Yongliang

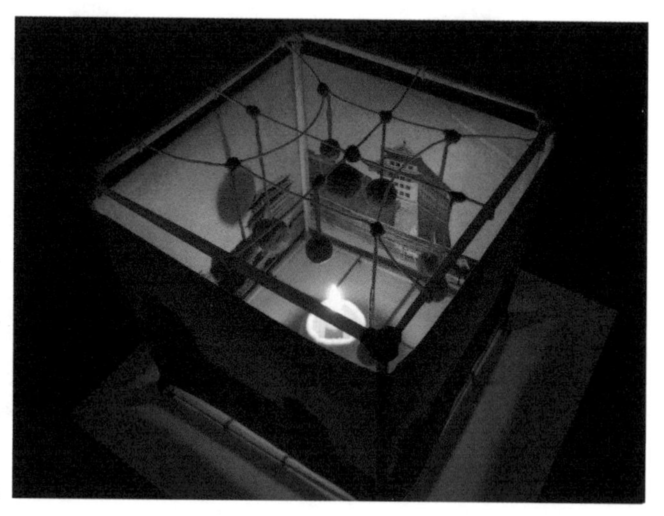

模型表达 :Ma Shuangzhi/Li Hongli/Xie Chenjuan

两校师生合影

颐和园西南
课程照片

2012.8~2013.1

01 场地调研 Site Survey

场地调研现状

设计课工作照

工作现场

02 设计合作 Design Cooperation

通过手工模型表达设计概念

现场指导与答辩

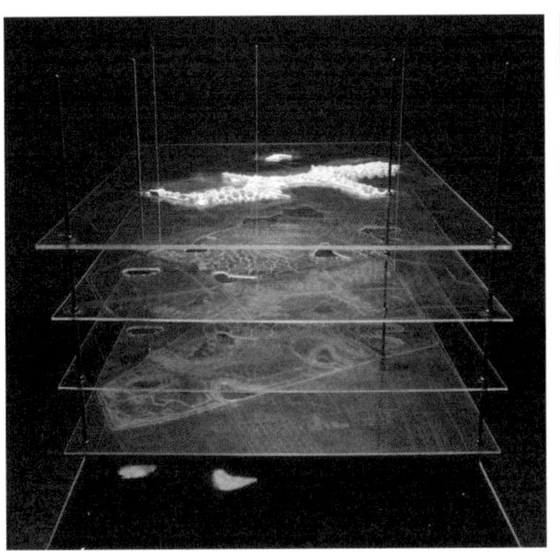

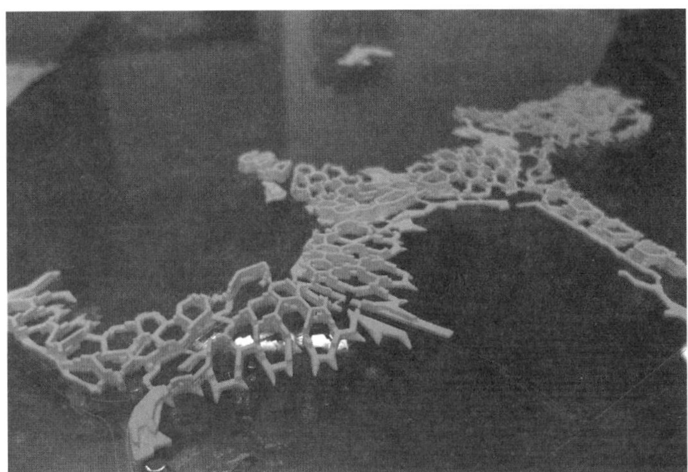

模型表达：Lv Hui

模型表达：Zhu Yijun

黄石铁山矿坑课程照片

2012.8~2013.1

01 场地调研 Site Survey

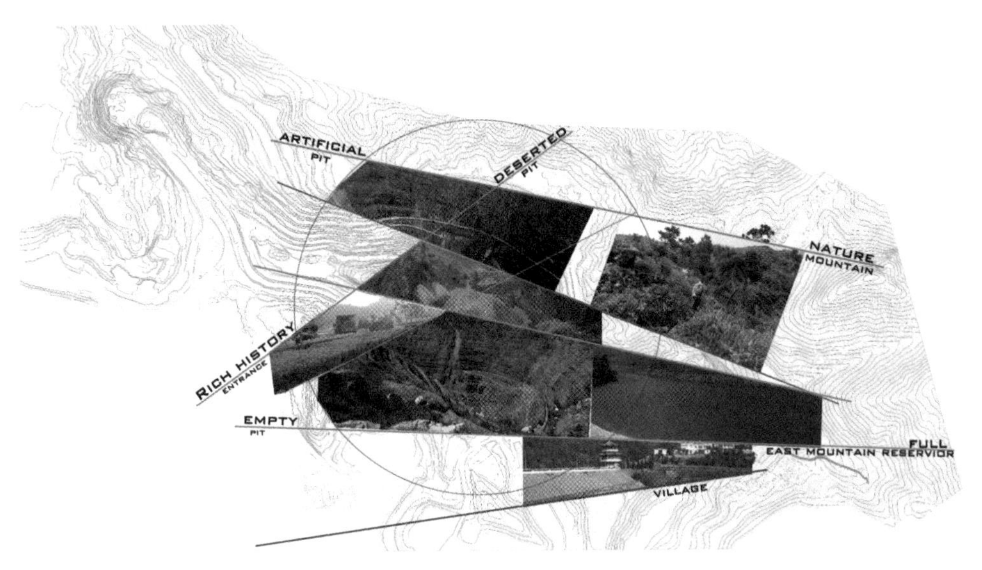

黄石场地调研

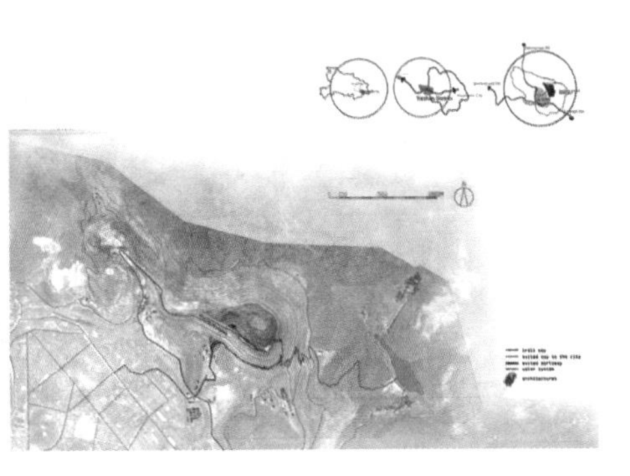

黄石矿坑现场调研

02 游学参观 Field Trip

巴塞罗那工作会议

考察巴塞罗那城市空间

联合 STUDIO 师生合影

THU/UPC EXHIBITION

印象北京 / 巴塞罗那展览内容

2012.9 ~ 2012.10

Preface

EXPLAINING BEIJING, REMEMBERING BARCELONA

The Department Of Landscape Architecture
School of Architecture, Tsinghua University, Beijing, China

EXPLAINING BARCELONA, REMEMBERING BEI-JING
ESCOLA TECNICA SUPERIOR D' ARQUITECTURA
DE BARCELONA. ETSAB (SUPERIOR TECHNICAL
SCHOOL OF ARCHITECTURE OF BARCELONA).
UNIVERSITAT POLITECNICA DE CATALUNYA.
UPC. (POLITECNICAL UNIVERSITY OF CATALUN-YA)
With the support of THE EMBASSY OF SPAIN IN
THE PEOPLE'S REPUBLIC OF CHINA.

PRESENTATION

Year by year the contact and exchange between professors and students of oriental and occidental universities is more frequent. However this exchange often does not deepen sufficiently enough into the different cultures and civilizations. It is more concentrated on concrete or technological aspects that are essentially neutral. This relation-ship limited to a narrow field implies that overcoming the limitations in the absence of a greater understanding can provide affliction to the exchan-ge. Maybe in the fields of Urbanism and Lands-cape, this necessity to broaden the knowledge of identities, and the intangibles that generate them, seems more evident. forms an unpublished cultural and artistic experience in which art is used as a mechanism to explain the intangible which configures the identity of a city. Identity is represented through the construction of artistic objects, abstractions, sensible interpretations and poetic models that seek to explain the poetry rather than the materialism of an urban reality. The constructions are inspired by an extraordinarily attentive reading of the reality. While "Explaining Barcelona" transmits to the Chinese visitor the identity of a European city, "Remembering Beijing" expresses a sensitive footprint produced by the Imperial city.

At the exposition "Explaining Barcelona. Remembering Beijing", a cultural activity between ETSAB/UPC-AS/THU, "Explaining Barcelona" exposed 20 poetic models accompanied by corresponding explanatory panels. In these panels the city's character is briefly introduced through the evoked fragments of the city and a brief description of the models and other interesting projects. "Remembering Beijing" exposed four digital's canvases, expressing the primitive perception of the huge city of Beijing.

The models have been crafted by the students of Project II The Medium Scale for the Master of Landscaping UPC and the degree in the subject Introduction to Landscaping. Subjects are taught at the School of Architecture of Barcelona. The responsible head and coordinator of the course is professor Miquel Vidal.

INTRODUCTION TO LANDSCAPING.

Relation of the students whose models are presented

01-The Raval district, as a historical reference and contemporany cultural crossroad.Multiculturality in-out .Eva Rosenova.Complex diversity. Boglarka Urban.Transition Marina Mazzamuto.
Forced Identities. Ana Quintana & Sergi Romero.
Feeling Raval. Elisenda Lurbes & Jesús Cuenca & Gina Sorolla.

02-The Eixample Cerda, essential identity of Barcelona.Discover. Boglarka UrbanExpansion. Variety . Eva Rosenova.Expansion. Perspectives?. Lucien Villier Eixample District contrast. Carla Casanova & Yasmin Castillo & Inma Santamargarita.Building your Own 22@ Pobe Nou. Paula Anasagasti & Anna Mallen.

03-The old villages annexed to Barcelona as an origin of identities. Sant Andreu. Gradients. Carla Casanova & Yasmin Castillo & Inma Santamargarita.Sant Andreu. Textile. V.Cagliani & M. Grañena & L. Verano.

04-The three big historic Parks of Barcelona as records of the history of the city.
Ciutadella Park. Memories garden. Anna Mallen & Sergi Romero.Ciutadella Park. Sonora models. Carla Casanova & Yasmin Castillo & Inma Santamargarita. Montjuic Park. Microlandscape. Paula Anasagasti & Ana Quintana.Montjuic Park. History layers. V.Cagliani & M. Grañena & L. Verano.Park Güell. Park- Boutique. Eva. Rosenova.

05-The urban infranstructures as a form to remake the city.Model Nus de la Trinitat. Estatic versus Dinamic. Johan Backman Model Nus Trinitat. Inner movement Eva Rosenova.

06-The poligons of new creation, an alternative reading of the Cerda plan.Besos Quarter. Massive Origami. Paula Anasagasti & Sergi Romero.
07-Barcelona's Sea Front as an account of subordination and opportunity.Barcelona seafront. The Forum. Domenec Llorca + Zhenguo.MASTER OF LANDSCAPING UPC PROJECT II THE MEDIUM SCALE
The poetic models explain through an artistic language the following identities:Space A.
1. The Raval district as an historical reference and a contemporary cultural crossroad.
2. The Cerda Pla and Eixample district. The essential identity of Barcelona.
3. The historical villages annexed to Barcelona. The origin of their identities.
Space B.
1. The three historical parks of Barcelona as records of the history of the city: Park Guell, Ciutadella and Montjuic.
2. An alternative interpretation of the Cerda Plan: The Besos neighborhood and the 22@ project.
3. The urban infrastructures as a way to remake the city. The "Nus de la Trinitat" Park.
4. Barcelona's seafront: an account of subordination and opportunity.
BRIEF HISTORY OF THE CITY OF MIRACLES: BARCELONA
The city of Barcelona's history is a paradigm through its clear development. As a city with a roman origins, Barcino, was a typical settlement with the forum in the center at the junction of the perpendicular axis cardus and decumanus. Its extension and localization corresponds today to the Gothic district.In the middle ages the city lived one of the most important moments of splendor. As a gateway to the sea of the Catalonian and Aragonese kingdoms, the city exerted commercial and military dominance in the Mediterranean. The Santa Maria del Mar cathedral, the Drassanes (Royal Shipyards) and the Palace of Tinell, are important gothic buildings that reveal the wealth and strength of Barcelona at this historical moment. The medieval city, in order to construct the second defending wall, incorporated an area of crops/orchards which soon introduced craft workshops and factories that later marked it with a certain marginality which has has become the contemporary RAVAL district.1714, and generally through the XVIII century, is a period of tragedy for the city in which the victory of the Bourbons erased the Catalan identity by prohibiting their language and institutions.The city was reborn in the nineteenth century with the industrial revolution, making it the first in Spain to generate dynamic, creative and highly tensio-ned industrial bourgeoisie and labor movements, mainly with anarcho-syndicalist roots. In the nineteenth century and the transition into the twentieth century, Barcelona defined a proper profile due to the Cerdá Plan, a Plan of Interior Reform and Extension in 1859 and Modernism. The CERDA plan as an urban area (which is later explained in more detail) has made Barcelona a leader of urban planning with its strong social roots beyond the anachronistic Beaux Arts. Modernisme was the style for the whole catalan society and covered all forms of arts and crafts: jewelry, graphic design, poetry, painting, music, etc. Barcelona is filled with countless traces, both humble and powerful ones, that show it. But the place where
Modernism formed the image of the city was in architecture, projecting unique buildings such as the Palau de la Musica Catalana (Palace of the Catalan Music) by Lluis Domenecc i Montaner and the Amatller's House with its blend of colors and dynamism that beautifuly recreate nature by Puig i Cdafalch. But above all stands Antoni Gaudi i Cornet, the man who certainly has given more glory and fame to the city,author of the Sagrada Familia, the Pedrera, the Batllo's House, etc. The city of Barcelona is in the center of a plain between two rivers and a mountain. It is the center of a crown of farming villages, Sant Andreu, Sant Marti, Gracia, Sants, etc., that eventually industrialized and expanded towards Barcelona, joining the city in 1897. These villages are the seeds of the precious diversity of a Barcelona of "Barcelonas".
Barcelona is not the capital of Spain and in order to progress must constantly reinvent itself and search for new territories and uses for the fields that are not yet urbanized or random neighboring parks. A new process of urbanization was the attempt to introduce a garden city to Barcelona through the creation of the residential Park Guell where Antoni Gaudi put all his creative capacity to shape the arid Mediterranean city into a lonely garden city. The project's failure opened the doors to the PARK GUELL fragment developed at the exposition.
The parks were favorable places for the new municipal lots. The first action was the Universal Exposition held in 1888, which readapted the whole project of Josep Fontsere to the Park of the CUIDADELLA (citadel park) which was to be given to the triumphant, nationalist bourgeoisie as a showcase to the world. In the history of the city, if there is one place that embodies all the lights and shadows, it would be the Montjuic mountain and its park Montjuic. It was an advanced, open work built for the Universal Exposition of 1929 and was kept hidden from 1939 to 1975, until the death of the dictator. The Fossar de la Pedrera (the Quarry common grave's memorial) inMontjuic park keeps in remembrance the victims from the facist repression in 1939-1952 through its architectural representation of light and darkness.
After the 1888 and 1929 exposition, the city planned another milestone in its development: the Olympic Games. In 1992 the Olympic Games, revived Montjuïc and the Montjuïc Park which became an Olympic space, as well as the whole city, opening it to the sea.
During the eighties Barcelona shined again in the world of urban design. The Barcelona model, in which public space is the engine of urban regeneration, is extended all over Europe. The city extended its design to all its fields: from the squares to the parks and from the urban designs to the infrastructures. The NUS DE LA TRINITAT is an example.
The constant validity and vitality of the Cerdà Plan demonstrates its ability to respond to different situations in different contexts. Thus in the firey developmental period of the fifties, in the polygons that settle between the neighborhoods and the river, the Cerdà block was reinterpreted. However the Besos district is an example of how high quality initial ideas fail in front of low elaborated designs. The 22@ urban project highlights once again the ability to adapt to a future program seeking to blend the technology industry and residences.
The WATERFRONT streches from the thriving medieval port to the commercial and leisurely spaces, creating a line along the seafront in Barcelona from which is the pulsation of life.

01-RAVAL

COMPLEX DIVERSITY

This work consists of two parts: the first one explains the width of the Raval, the Boqueria market, which is the most important specialized fruit market of in Barcelona, the ethnic trade, and the diverse ethnicities which are customarily found in the Raval. The second part of the model shows the Raval as a puzzle, a complex reality that can be disorganized and reorganized over time.
Boglarka Urban.

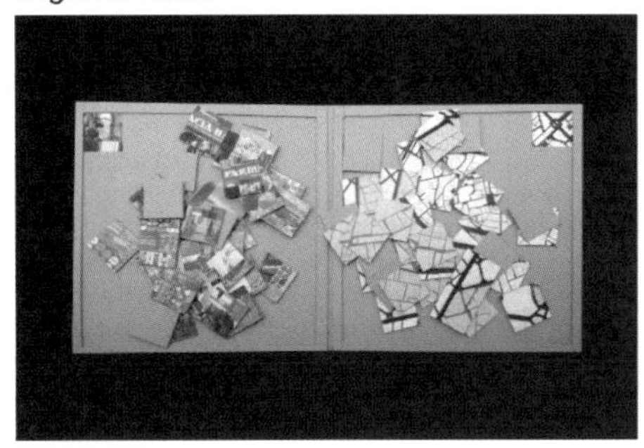

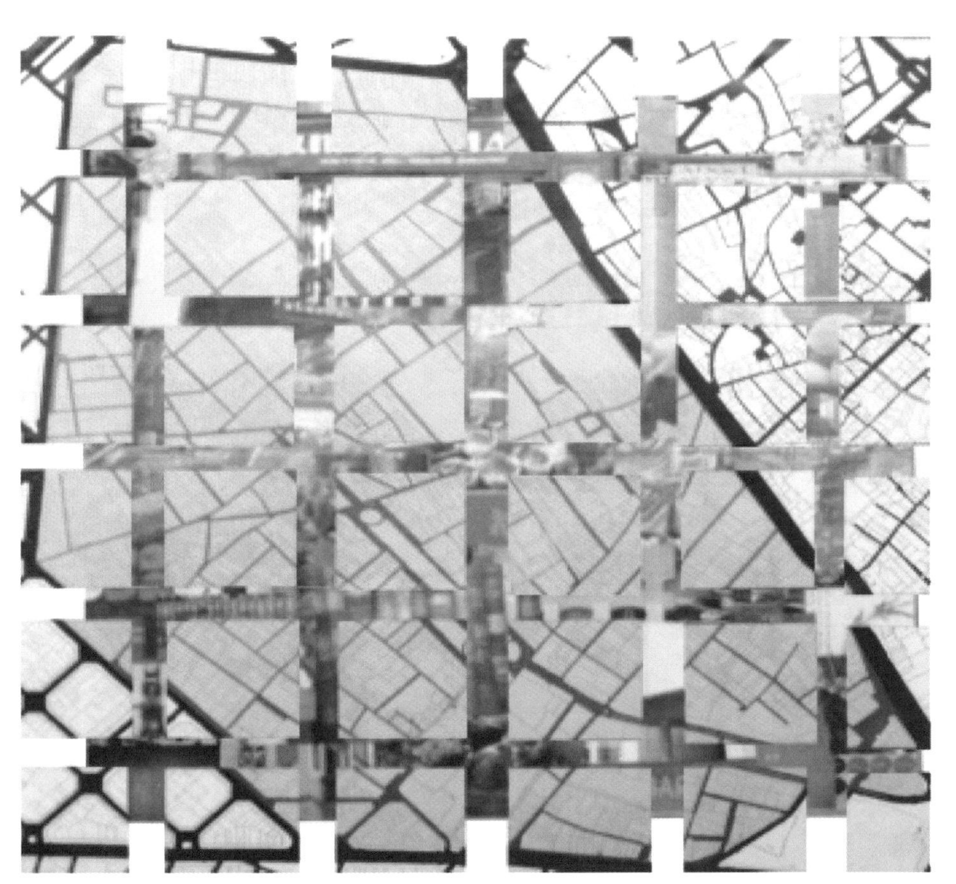

THE RAVAL DISTRICT, AS A HISTORICAL REFERENCE AND CONTEMPORARY CULTURAL CROSSROAD.

Miquel Vidal.

Between 1770 and 1840 a definitive industrialization was produced at the Raval district. During the second half of the 1700s, new streets with factories and housing for workers made their presence. Guild houses disappeared and were subdivided into many rental housing units to accommodate the many peasants fleeing the famine of the field (agriculture crisis of 1765-1766). The factory workers were left to live in the Raval, close to their working place. This neighborhood became the densest in Europe and took up the last buildable square feet of the city.

The first voices to reclaim the improvement of the neighborhood arose in the thirties, during the Second Republic in 1931-1936, with the proposals of the GATCPAC architects. Since the eighties, the Administration has promoted a strong policy of reform and rehabilitation of housing, openness of spaces and creation of facilities for the community, at a city level.

In the Raval district of Barcelona, models of poetic discourse through art objects are based on: 1) The multiculturalism due to its ethnic diversity, 2) The strong existing contrast between the gradient grid based on the multicultural environment and the presence of new, high-quality buildings with cutting-edge architecture (such as the Center of Contemporary Culture, the Museum of Contemporary Art in Barcelona by Richard Meier and the Film Library of Catalonia by Josep Lluis Mateo) and 3) The changing character of the new public spaces generated by "Empty health" and defined by the program for improvement in the Raval district.

01-RAVAL *FEELING RAVAL*

Raval neighbourhood is located in the historical centre of Barcelona. Is one of the oldest neighbourhood of the city, together with Gothic and Borne. A destacable characteristics is the massive arrival of immigration through the years, due to the deterioration and antiquity of the buildings mainly.
Now a days the city council is trying to open the neighbourhood introducing new equipments offering several services, such as recreative or cultural, to move nearer the rest of Barcelona's citizens and not to create a ghetto.
The model tries to reflex what we consider is happening at Raval every day, where citizens uses this news equipments and some of the olds, for it's momentary use without having any repercussion on the life neighbourhood and his economy.
Starting from a white linen we drew the streets with ropes as the architectures use to build the streets in the ancient. The colours are trying to reflex people density coming from other parts of the city or foreigner. To show the activity of the buildings expressed before, we decided to elevate the buildings by means of balloons those equipments that didn't belong to the real necessities of the neighbourhoods that suffer them every day.
The model has a claiming background to make evident that the political necessities and the ones coming from the citizens almost never goes hand by hand.
Jesús Cuenca + Elisenda Lurbes + Gina Sorolla

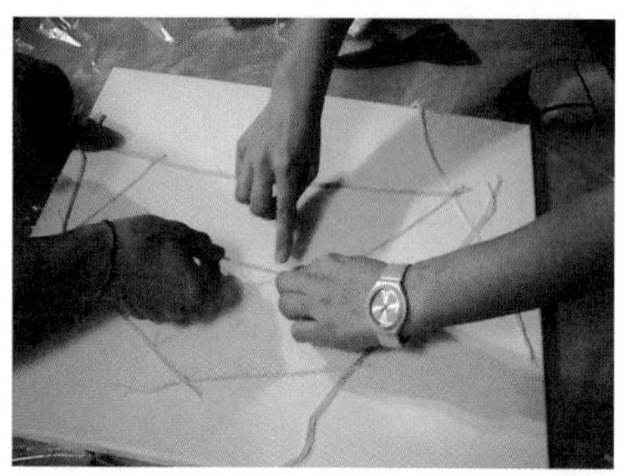

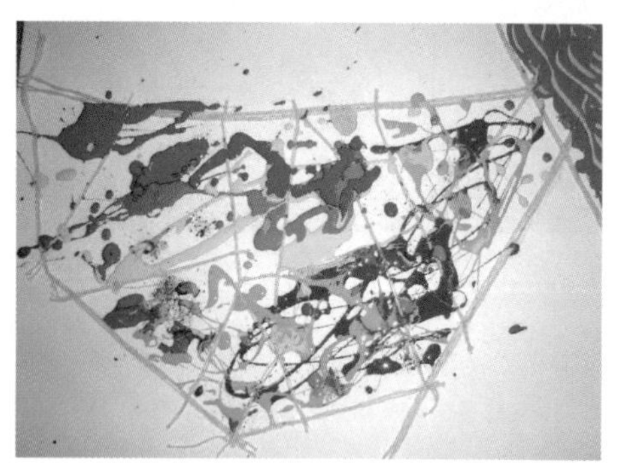

01-RAVAL *FORCED IDENTITIES*

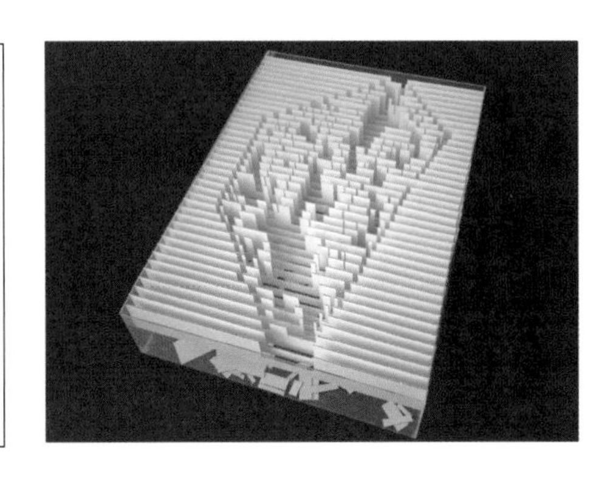

The area of study is the Raval district, a dense and impermeable neighborhood with very clear limits. In the last years it had been submitted to a deep renovation with the purpose of swelling and introducing new uses and open spaces in the area. The new interventions get into contrast with the essence of the district and generate new dynamics around it. The sensation of density that the area transmits when you walk through it and the opposite sensations that the new interventions with a very different morphology promote, is our starting point. A sequence sections recreate the urban fabric before the emptying. The trim that is made to this structure explains the gradient of voids and the connections between them. Different types of fissures appear: narrow and shallow fissures evoking the characteristics streets of the old town and some others broad and deep fissures like a symbol of the sharpness of the recent interventions. These new and broad fissures that appear, colonize the pre existing structure and degrade the actual model. The already explained sequences have been overlapped with the extracted volume emphasizing the proportion of demolition against the actual urban fabric. The magnitude of the new surgery processes, new uses and needs that had been implemented are somehow reflected in this model.
Ana Quintana+Sergi Romero

01-RAVAL *SHAPE AND TEXTURES*

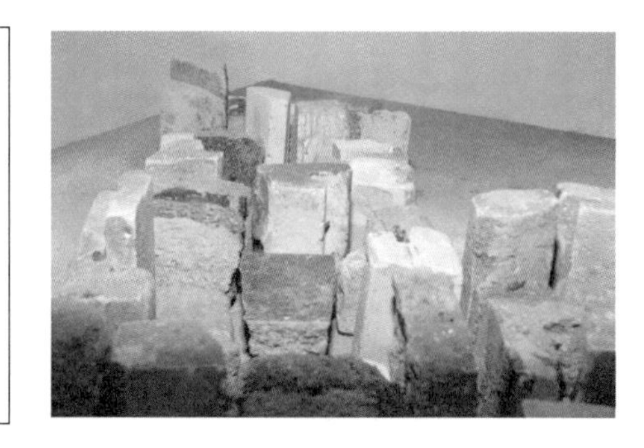

El Raval is one of the areas inf Barcelona's historic center and it is noted for being a poor and multicultural area. The developed project consists of the reconstruction of part of the area with certain degree of abstraction in order to reflect in the urban fabric the textures and the shape which characterize it.

The chosen line of work, taking into account that the objective of the project is reflecting the sensations this particular area of the city evoked in us, is to obtain both the shape and texture of El Raval with debris from the project area itself.

Consequently, the output generated is, in itself, part of El Raval, which explains both the shape of the streets and the urban fabric as the feelings you get when you walk through its streets.

You could say that the best way to convey the feelings of El Rava is by itself.

Aitor Arconada + Diego Brenes + Xavier Garcia

01-RAVAL
MULTI CULTURALISM IN-OUT

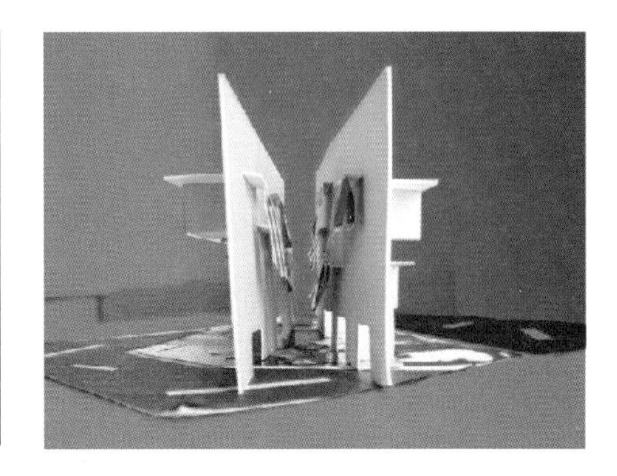

This model seeks to explain the different nationalities living in the Raval through the use of flags that, relate them to private areas and areas exposed to the street. The street is a meeting place and a place of contrast, where the external aspects of each ethnic group, such as color, clothes, jewelry and religious symbols are presented in a contrasting form. Whereas In the privacy of every home everything is uniform and undistinguished by ethnic differences. If anything could distinguish the inner-home space, it would be occasional social or economic distinctions. It is also a place for practicing religion and traditions.
Eva Rosenova.

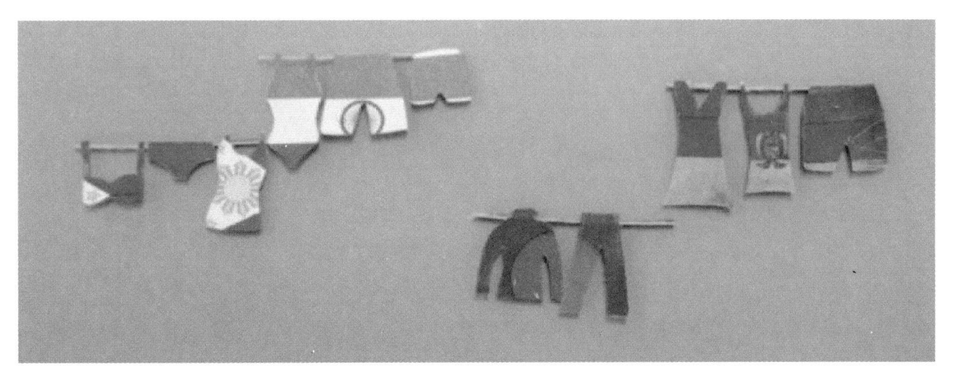

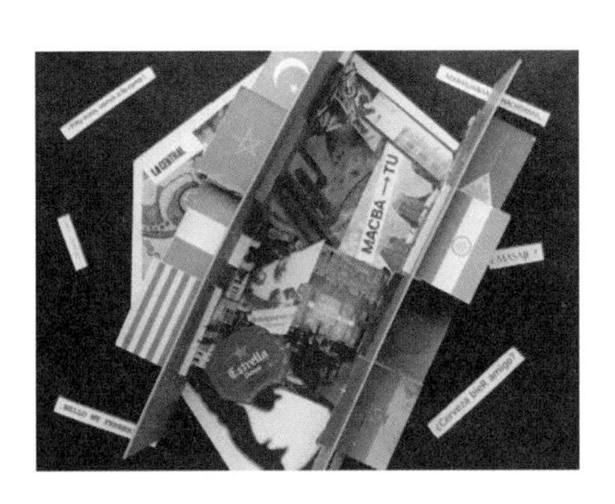

01-RAVAL
TRANSITION IDENTITIES

This proposal refers to the origin of the marginalization which today is manifested in the Raval. The walls had lost their military value and had become the dividing lines between the center and the periphery of the city long before they were demolished. This fact determined the marginality of the neighborhood and the progressive deterioration of its buildings.
Marina Mazzamuto.

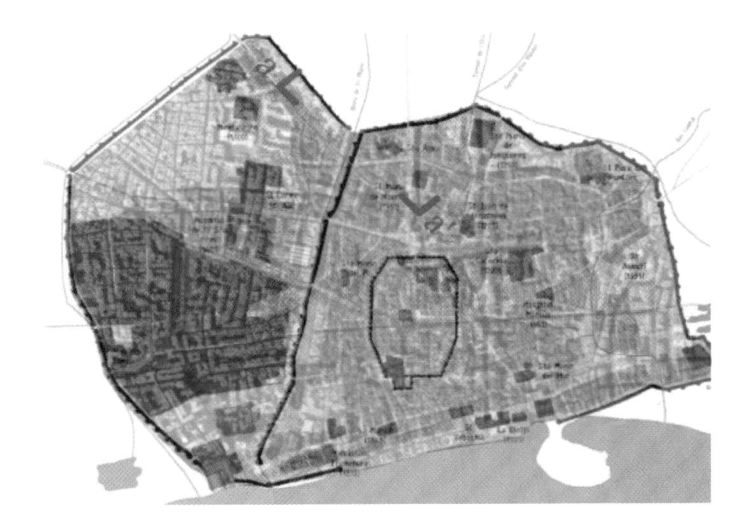

02-EIXAMPLE DISTRICT BUILDING YOUR OWN 22@-POBLE NOU

The proposal aims for emulating the buiding development taken place in 22@ and Poble Nou. As a means of understanding the current situation of the area. Through modeling the scenery the bystander becomes the actor and faces the contrast,tension and clash felt along the district.
The idea of Time as a main part of the process is introduced by getting in touch with the model itself.
Undeniable crucial role is hold by the Time in the streetlandscape of the area, its presence is sensed, even, through the numerous blank spaces and plots filling the blocks.
A board, pieces and instructions are shaped while trying toportray the scenery as well as introducing the spectator asa taking part role. This way the observer becomes part of the proposal, while standing as a main piece, activates the process and brings out the Time concept.
'Reflect through Action' is the means to achieve an understanding of the workings of the city, its urban development as well as emulating some of the various circumstances holded in Poble Nou and its surroundings, such as the remainigs of the old agricultural layout, the lingerings of the industrial function, Cerd·'s grid or the last projects carried out by the 22@...
Paula Anasagasti + Anna Mallen

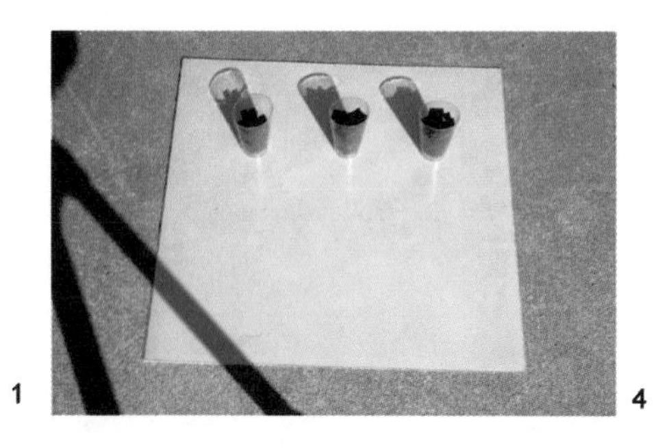

1

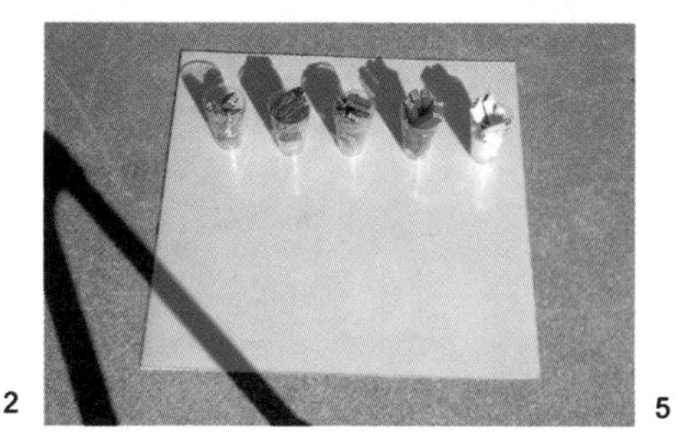

2

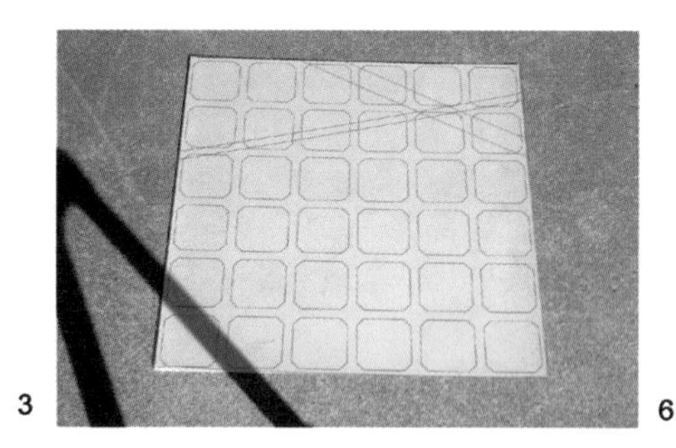

3

4

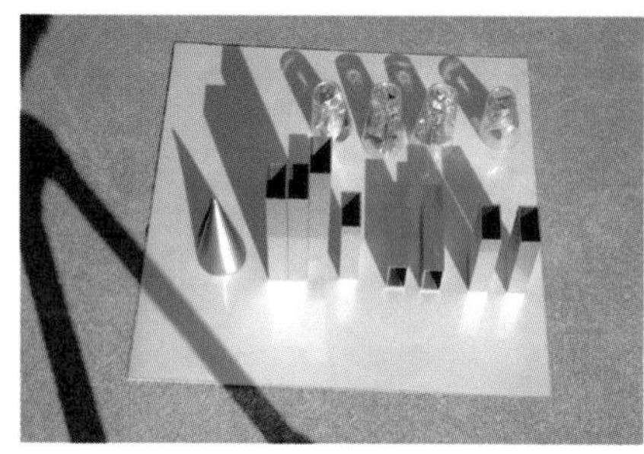

5

6

PLAYING ELEMENTS

THE EIXAMPLE CERDA AN ESSENTIAL IDENTITY OF BARCELONA. Miquel Vidal

The 1859 Internal Reform Project and Extension Project in Barcelona by Ildefons Cerdà, is based on a large network of transverse and perpendicular streets. All of these axes are uniform, except for two oblique overlapping streets, the Diagonal and the Meridiana, and the Gran Via de les Corts Catalanes which crosses through the Glories shopping center. The point of intersection of these important thoroughfares was the great communications center of the Eixample district, where a big square was planned to be constructed Glories shopping center.

The point of intersection of these important thoroughfares was the great communications center of the Eixample district, where a big square was planned to be constructedGlories shopping center. With great precision, it predicted the uniform distribution of service areas such as markets, community centers, churches, and large district parks.

The city blocks, with dimensions of approximately 121m x 121m, were not exactly square. For visibility reasons the corners of the quadrilateral were cut on an angle of 45°, which gives a unique functional and landscape quality to the Eixample district. In each block it was permitted to construct only one or two sides, and the remaining space was left as a garden for the neighbors. Cerdà established this specific form of the blocks as he considered that the health of citizens depended on a good orientation of dwellings. Cerdà is also the author of the 1867 General Theory of Urbanization, a pioneering work of urban science.

"My whole fortune, all my credit, my time, all my comforts, all my affection, and even my personal consideration in society, (is based on) the idea of urbanization" The General Theory of Urbanization.

The elaborate works focus on one of the great virtues of the Cerdà Plan: the versatility and ability to support any activity.Every activity is deeply intertwined commercial, cultural, recreational, residential and socially. This exemplifies one of the best examples in the world of a complex and fragmented city. In the poetic models it shows that behind an apparent uniformity lies a widely diverse city.

02-EIXAMPLE DISTRICT
CONTRAST

FROM INDUSTRIAL TO TECHNOLOGICAL SUBURB

Contrast of an urban and rural fabric. That rural tissue was fractionated by overlaying mesh Plan Cerdá apples.
In the evolution of the site where the buildings have adapted various positions, from adapting to the container (the apple of 113mts x 113mts) to differ from these partially or fully.
In our demo we show how gradients are behaving as coforme architecture of an industrial advances (materials typical of the area) to a polygon own technology architecture area 22 @.

C.Casanova + Y.Castillo + I.Santamargarita

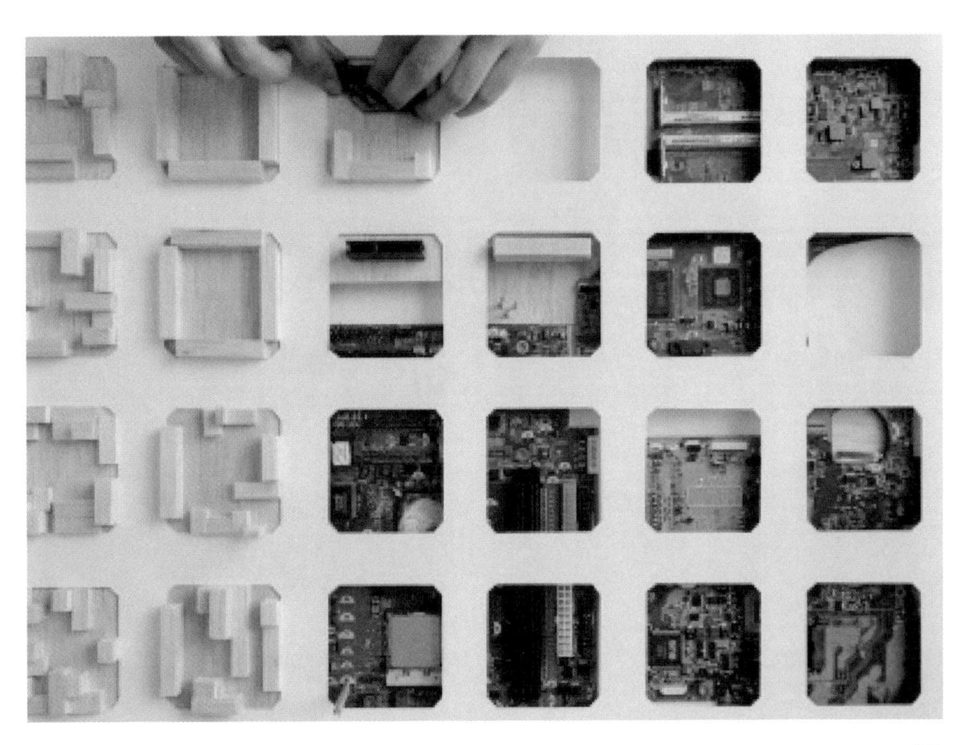

02-EXPANSION/ ENLARGEMENT
DISCOVER

This plan expresses the same concept through a wide variety of uses in architecture, different environments and shops that all characterize the Eixample district. The double scale refers to a large scale that represents the repeated blocks which, in the eye of the visitor, form a monotonous succession of buildings. On a similarly monotonous but smaller and more personal level, is the scale representing trade and business that occurs on the terraces. The Eixample district's streets' and terraces' spatial relationship was proposed to be 5 meter wide sidewalks and 10 meter wide.
Boglarka Urban

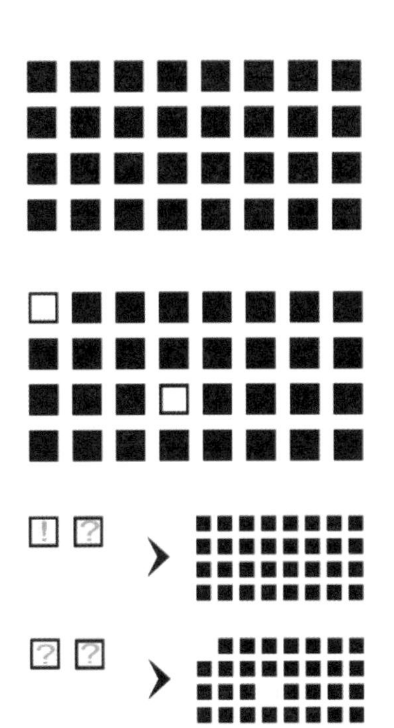

THE EIXAMPLE CERDA` AN ESSENTIAL IDENTITY OF BARCELONA. Miquel Vidal.

The 1859 Internal Reform Project and Extension Project in Barcelona by Ildefons Cerdà, is based on a large network of transverse and perpendicular streets. All of these axes are uniform, except for two oblique overlapping streets, the Diagonal and the Meridiana, and the Gran Via de les Corts Catalanes which crosses through the Glories shopping center. The point of intersection of these important thoroughfares was the great communications center of the Eixample district, where a big square was planned to be constructedGlories shopping center. With great precision, it predicted the uniform distribution of service areas such as markets, community centers, churches, and large district parks.

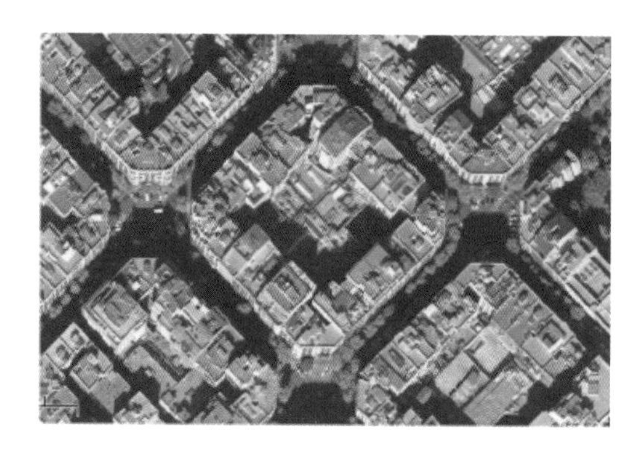

02-EXPANSION/ ENLARGEMENT *PERSPECTIVE?*

One of the most common and perhaps less important criticisms is the absence of "monumentality" in the Cerdà Project in comparison to Antoni Rovira i Trias's project chosen for the city in which the influence of Haussmann's renovation of Paris shows a strong monumentality. The artistic proposal was based on the intention to interrelate, in a participatory way, the existing monuments with the basic perception of the Cerdà Plan through the 20 meter wide streets.
Lucien Villiers.

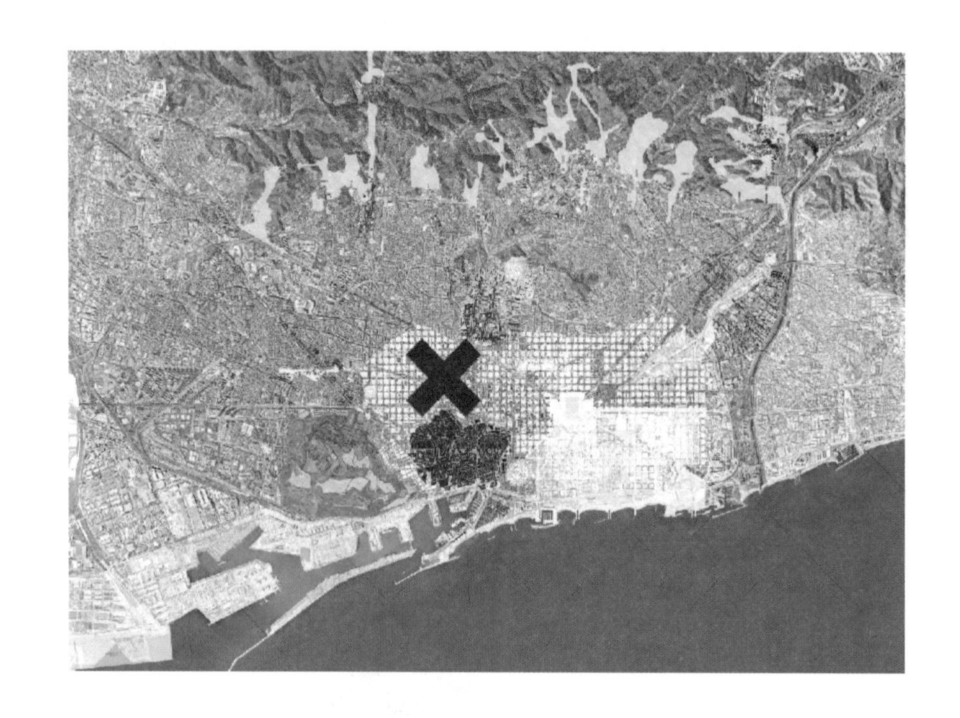

02-EXPANSION/ ENLARGEMENT *VARIETY*

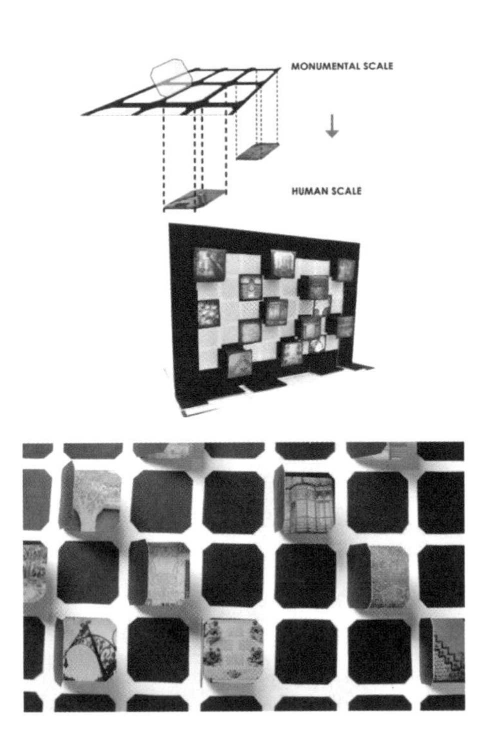

2. THE EIXAMPLE CERDA. ESSENTIAL IDENTITY OF BARCELONA. Miquel Vidal.

Details_human scale

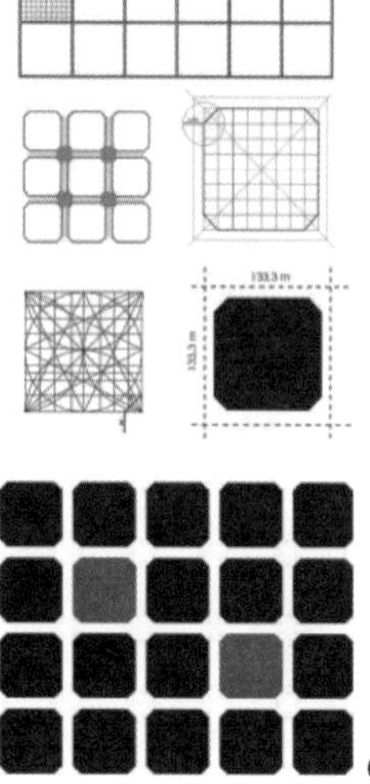

Geometry_monumental scale

03-SANT ANDREU
GRADIENTS

We took a tour from the ends of Sant Andreu where even the central area where two layers are evident:
A layer showing gradients from architecture to an architecture with identity apart from it, to another level.
A second layer showing social gradients where the closer to the center increased activity is generated and is lost as it approaches the ends.
The pieces that represent the architecture of the old town are more permeable while parts are closed perimeter evident the contrast between the two architectural interventions

C.Casanova + Y.Castillo + I.Santamargarita

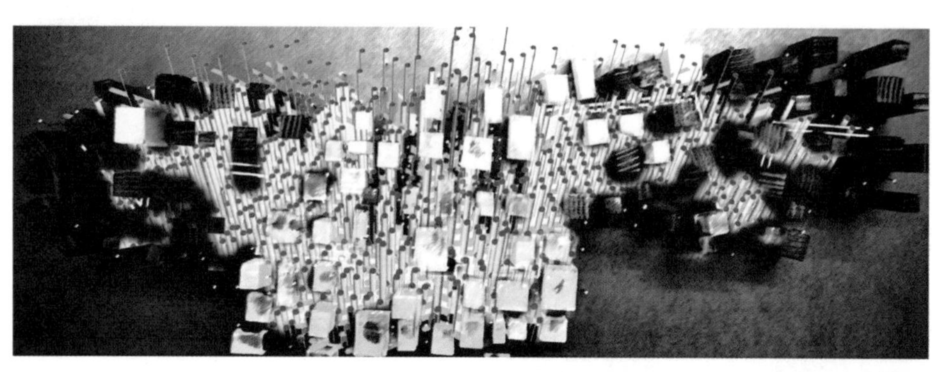

THE OLD VILLAGES ANNEXED TO BARCELONA. THE ORIGIN OF THEIR IDENTITIES. Miquel Vidal
The "Barcelonas", the ones referred to by Vazquez Montalban, have their origin in the ancient villages of Barcelona's plains: Sant Andreu, Sant Marti, Gracia, Sants, etc.. The first script written as a reference to the village Sant Andreu de Palomar dates back to 992. The village's main source of wealth was irrigated agriculture. Over the years, urbanization spread into the small streets around the Carrer Gran, the main axis and an ancient Roman road. In the late nineteenth century the following major companies were already installed in Sant Andreu: Fabra i Coats, La Maquinista (The Machinist) and the National Manufacturing of colorants.

The villages, annexed to Barcelona in 1897, maintained peripheral characteristics of numerous brown fields or vague terrain until recently. Now, it is finally being silted with increasing road networks and the arrival of plans for large interchanges such as the TGV station at the Sagrera in the Sant Andreu district which is already under construction.
In one of the proposals presented, increasing the tension in these parts of society is expressed as a transposition of physical stress like on a rubber band. The second proposal, makes a reference to the characterization of the district through the textile industry.

03-SANT ANDREU
TEXTILE

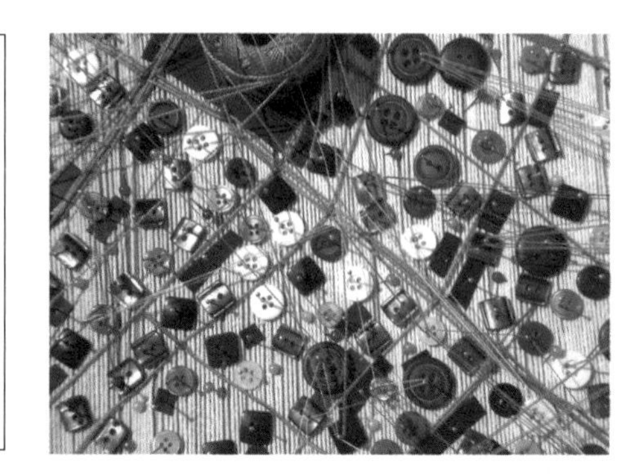

The model intents to express the influence of the industrialization over the historic agricultural grid of Saint Andreu district.
This is manifested with the creation of industrial nudes which act like focal points of extended social influence.
The physical character of the industrial localizations is expressed with buttons, whereas the social influence through threads that interlock basic elements of the textile industry which have transformed the old agricultural village.
The color is important as it reflects the chromatic pallet of the textile production. The predominance of blue is a memory- honor to the blue uniforms of the textile workers.
V.Cagliani + M.Grañena + L.Verano

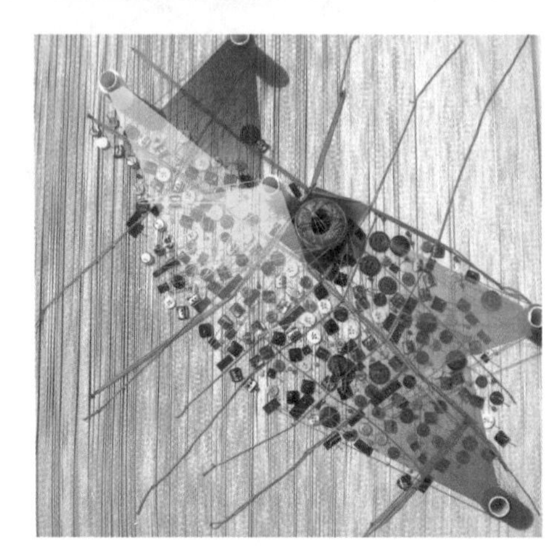

04-CIUTADE-LLA PARK *SONORA MODEL*

The space Ciutadella Park becomes historically an area a green walled content. Although it is located strategically in Barcelona and has hits that connect with the urban outside, inside there is a definite idea of the route when we left the park entrance shaft.
In another of his hits that mark the Meridiana Avenue into the park we have a limited visual and physically interrupted by the back of the cascade, seems to give back to the city. Once we are inside the park no sense of direction to the outside because the visuals are blocked by vegetation, interrupted only at points of access. Just as the sounds of the city that are generated abroad are being replaced by the sounds of the park.
To perform the analysis of sensations, we have focused on two parameters that affect reading this landscape, lights / shadows and sounds. With each of these parameters have generated ideograms, which have helped us to identify points of transition between different areas.
The analysis of the park begins at the Arc de Triomphe, located in the 'Passeig de Lluis Company', although this does not belong physically to the park, it is not understood without them. The ride itself is a hotspot while a meeting place. From here, we have been identifying different sensations throughout the park, associated to different areas.
Fraccionamos the tour into segments that follow sensory patterns, responding to the aforementioned parameters, lights / shadows and sounds.
To facilitate the creation of the ideogram, the tour is simplified in a straight line.
In each of the divided fragments are evaluated parameters set by differentiating the two central part of the route and both sides of this. Representing in the upper and lower line respectively.
Once the ideogram translates to a line that create different turning points, which will be to mark the change of representation systems. The representation of this conceptualization is materialized by a number of slots where the transition produce textures and density changes.

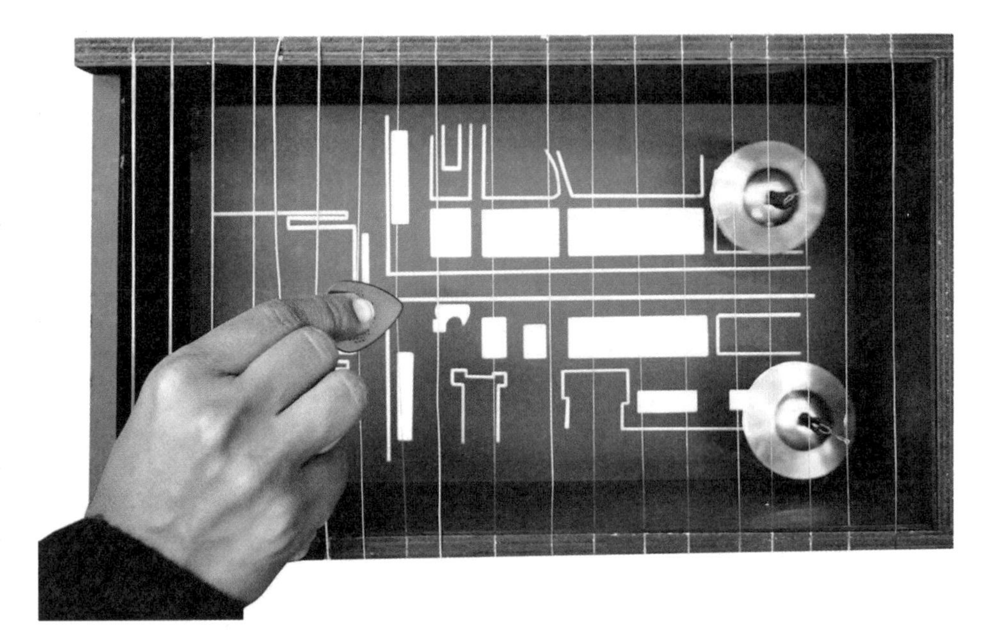

The different sensations are evaluated using a scale of values ranging from 0 to 10, where:

DIAGRAM LIGHT / SHADOW
0 _ maximum light no shadows
10 _ minimum brightness lot of shadows

CHART SOUNDS
0 Maximum number of city sounds
10 Maximum number of park sounds

Imma Santamargarita + Yasmin Castillo + Carla Cananova

04-CIUTADELLA PARK
MEMORIES GARDEN

The perception of time in the "Parc de la Ciutadella" is a constant fact when it is visited for who knows the city of Barcelona.
On the one hand we recognise series of elements strongly linked to the urban grid such as axes, diagonals, sculptures and fountains and on the other hand a few elements are inconsistent, without spatial continuity, like barriers, traces of old limits and perimeter buildings, which emphasize a feeling that something is not working out.
The model represents a retrospective of the main phases which it has moved this landmark green space from Barcelona. Trough different layers, transparent structures show us the main axes and characteristics of the park plot in several historical periods, from the fortress until the future project breaking walls to the waterfront.
Not less important is the park's memory, so the model especially emphasizes the elements which had coexisted along various phases and the ones that nowadays persist. Several threads link those elements, creating a framework that shows importance of permanence or transience on different parts along its history.
An analytical and conceptual interpretation is represented in this nano-landscape, which wants to show the park's structure through a backward glance as well as in the near future.
Anna Mallén+Sergi Romero

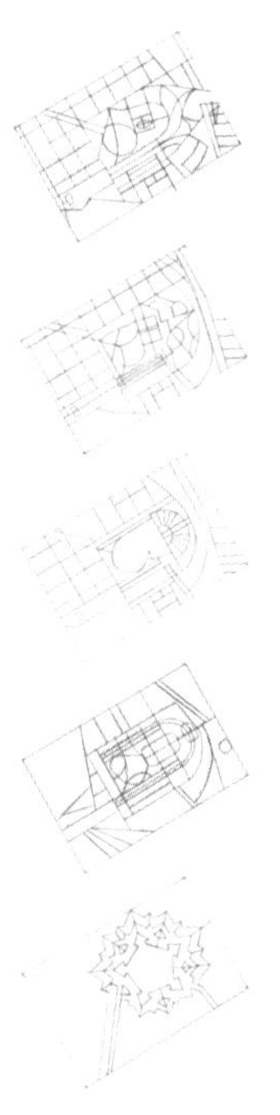

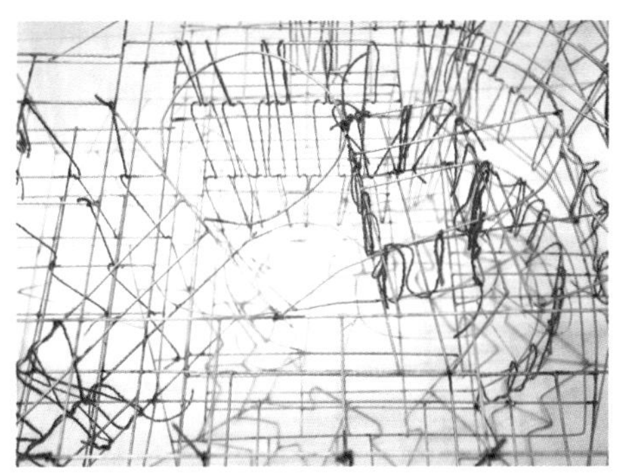

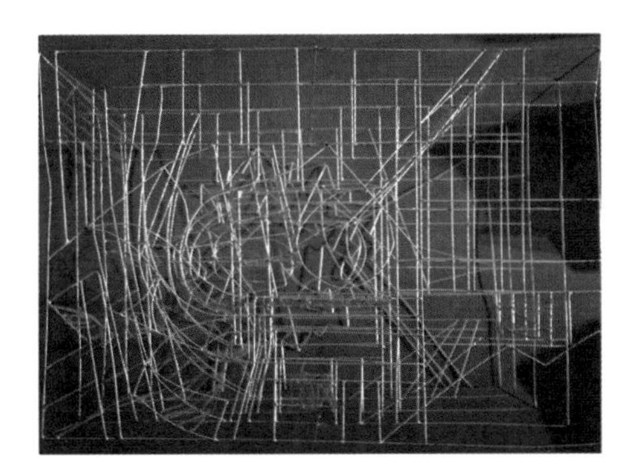

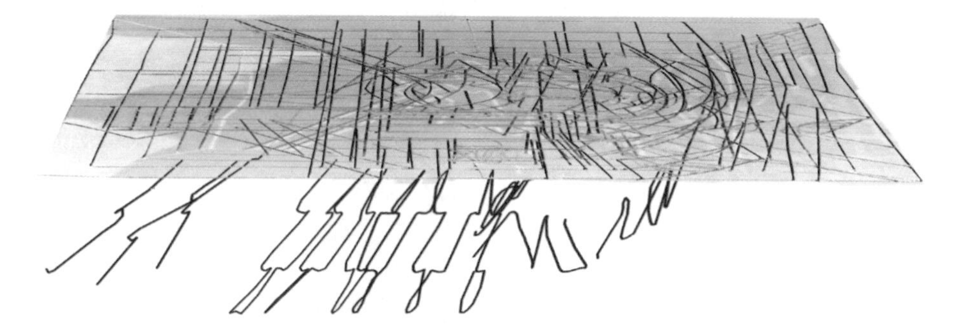

THE THREE BIG HISTORIC PARKS OF BARCELONA: PARK GUELL, CIUDADELLA PARK AND MONTJUIC AS RECORDS OF THE HISTORY OF THE CITY. Miquel Vidal.

The Ciutadella, which takes its name from the former military Citadel was the first major urban park in Barcelona and its disigner Josep Mestres i Fontsere, perfectly combined in his proposal the role of the hygenic urban park with the slogan "parks are for the city, like lungs are for the human body." He used Jean Charles Alphand's Parisian guidelines.

The park's structure fit with Cerdà's guidelines and the railway line between Barcelona-Mataro. Inside the park is a geometric garden and different landscape. Its design and its ideology was denaturalized By becoming part of the 1888 exposition. The park remains in a long process of decline and introduction of new uses such as the zoo in 1892.

The models express in a plastic form the park's evolution as if it were a palimpsest showing faint, original traces of the coexisting 1880 Competition and 1888 Exhibition which sharply are intertwined with an enormous zoo.

04-PARK GÜELL
PARK-BOUTIQUE

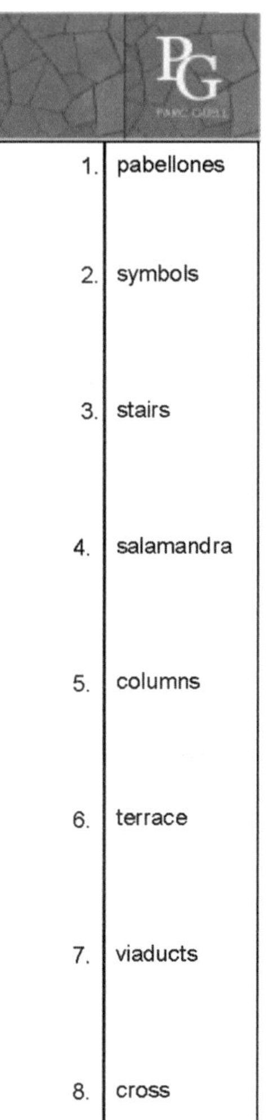

PG
PARC GÜELL
1. pabellones
2. symbols
3. stairs
4. salamandra
5. columns
6. terrace
7. viaducts
8. cross

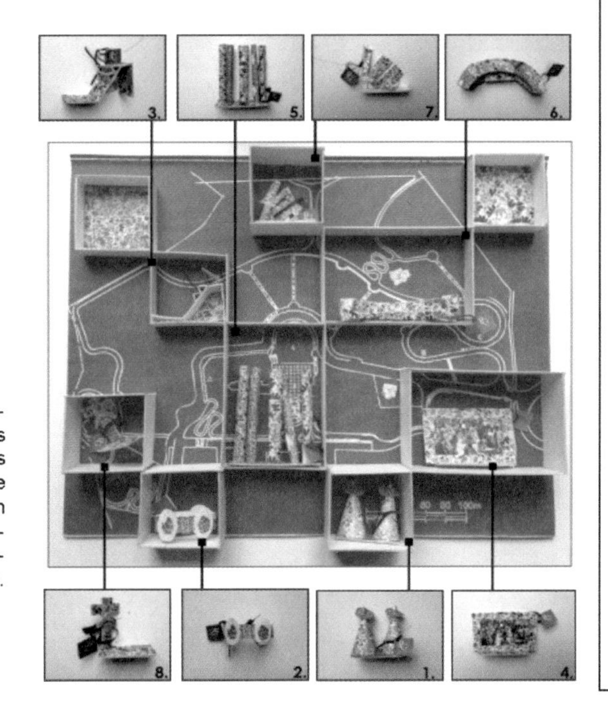

Park Güell is undoubtedly one of the greatest monuments in the city attracting hundreds of visitors daily. Organized tours only visit and take photos at specific points of attraction which leaves the park's overall structure as a complex transition between city architecture and nature out of context. The proposal focuses on explaining this phenomenon and, linking the park to a souvenir shop. Eva Rosenova.

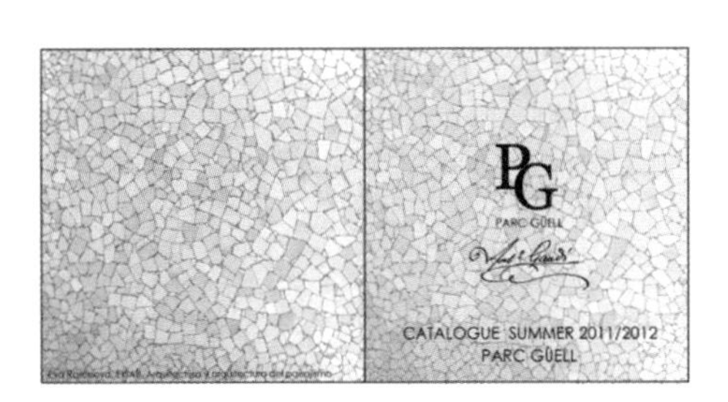

04-THE THREE BIG HISTORIC PARKS OF BARCELONA AS RECORDS OF THE HISTORY OF THE CITY: PARK GUELL, CIUDADELLA PARK AND MONTJUIC.

Miquel Vidal.

Park Güell originated in Count Güell's project to build a garden city on a topographically difficult farm with poor vegetation which he acquired in 1985. Gaudi first designed the urbanization of this space combining architecture and nature by blending organic forms and structures. Nature gradually imposes over architecture as you enter the park at either the pavilion entrance or the Hispostila Chamber (planned as a market) and large plaza. The garden city project was abandoned in 1914 and was opened as a public park in 1922.

The labor carried out in Park Güell affects the park's "unbridled consumerism" which has become a universal icon in tourism.

04-MONTJUIC *MICRO LANDSCAPE*

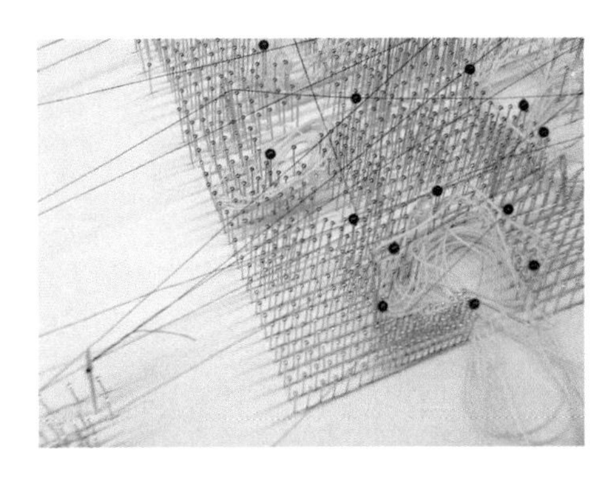

Is Montjuic just a part of Barcelona?
If we look back, a place that now is considered just one of the parks of the city, used to be a place that has established an inevitable and variable relationship with the city that is situated at its foots.

The quarries are used as elements that allow to describe the hill, from its geological formation, through its history, until its current use. In other words, as a conductive thread which articulates the entire argument and allows to develop a relationship between the objective data and the spacial sensations that characterized Montjuic.

The model aims to generate a micro landscape; Montjuic is only the starting point.
To achieve that objective, is established an abstraction process. In first place, some parameters are set to describe the group of quarries and its relationship with the city of Barcelona. Through the redefinition of these data it would be possible to set up the new horizon.
Precisely, the model virtue lies on the multiple interpretation which may have, not just like a model which represent an abstract Montjuic but an aim which recreate an space with endless readings.
Are maybe the pins trees? Are maybe just lampposts? What would be the threads?
Paula Anasagasti + Ana Quintana

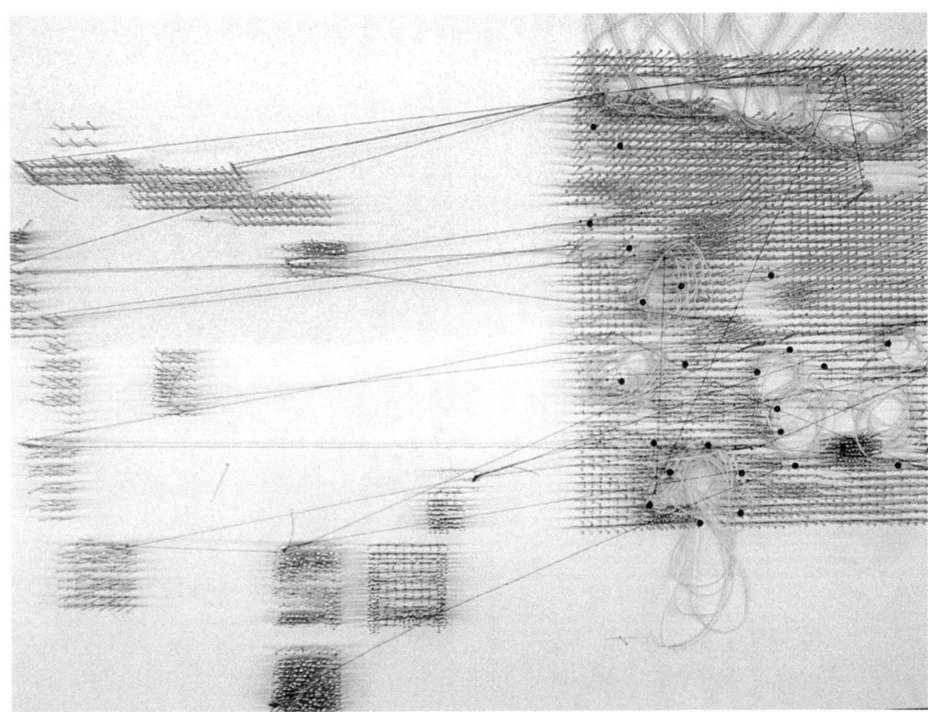

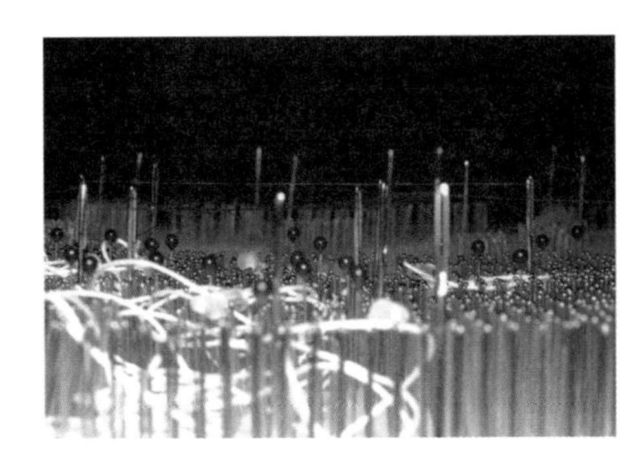

MATRIZ DE TRABAJO

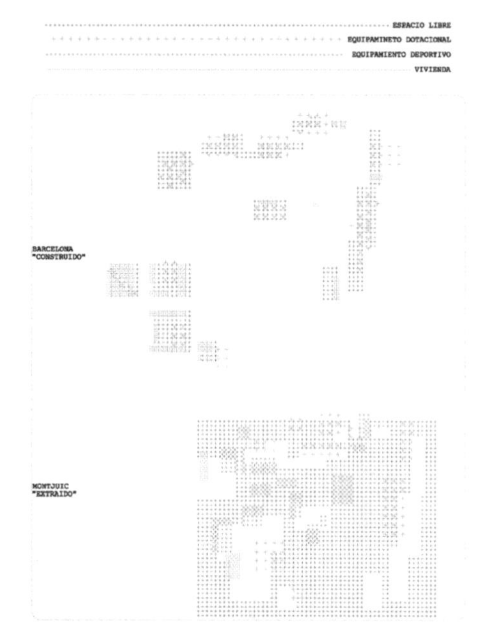

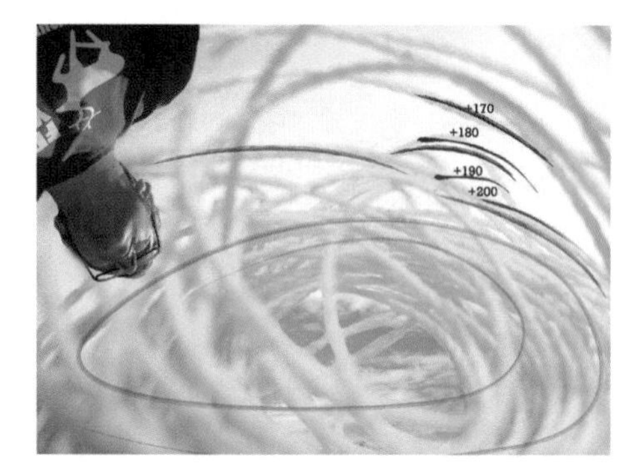

THE THREE BIG HISTORIC PARKS OF BARCELONA AS RECORDS OF THE HISTORY OF THE CITY: PARK GUELL, CIUDADELLA PARK AND MONTJUIC. Miquel Vidal

Montjuï c Park and the mountain on which it sits, represent the history of Barcelona, crowned by the Montjuï c castle which was designed by Joris Prosper Van Verboom in 1716 and submited to the city for its black past. The Montjuï c Park received its identity from the Universal Exhibition in 1929, as an urban park. It has a rugged topography which has always hampered a complex and diversified structure beyond the basic road designed by Josep Amargos in 1914 during its urbanization in 1894. During the exposition Maria Cristina Avenue linked the park to the city defining the broad perspective of the National Palace and framing the Magic Fountains. The dispersed architectural monuments such as Puig i Cadafalch's late modernistic pavillions and Mies van de Rohe's moderon pavillion are also adjusted to the topography. The park became a modern, Olympic stage in 1992 and a pool of cultural activities through the Mies van de Rohe pavilion, Caixaforum, Arata Isozaki, Joan Miro Foundation, Josep Lluis Sert, etc.

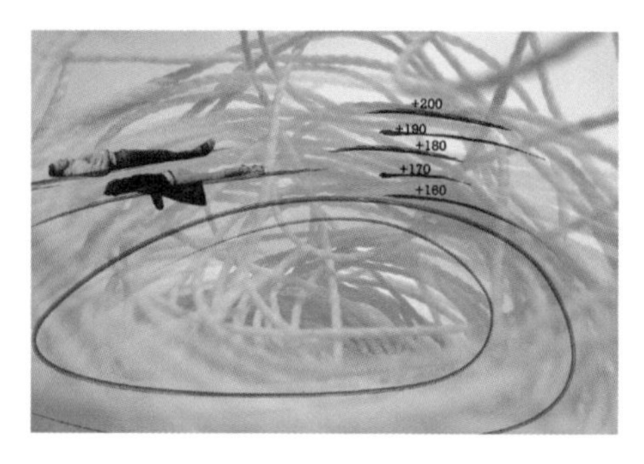

04-MONTJUIC *HISTORY LAYERS*

The study area includes Montjuï c mountain-city relationship.The city has grown up at his feet and the mountain has always been a strategic place from which to defend the city, so since ancient times has been a fortress on top. In 1751 they built the present castle, coincides with the dark stages of our political history and loss of freedoms.
In 1842 was bombed by the oppressive forces of the Spanish crown, later to suppress popular revolutions and finally the castle was used as a prison and place of shooting of political prisoners of the dictatorship of Franco (1939-1975)
Monyjuï c was also an economic symbol . The promotion and urban regeneration during the Universal Exposition of 1929 generated different pavilions and gardens that will be used later by the population of Barcelona. The sources and modern lighting gave a spectacular and entertainment center at the mountain.
The celebration of the Olympic Games in 1992 was another impulso regenerador , construyendo y restaurabdo nuevas i viejas zonas de recreo.
The representation of the various experiences that we had the mountain inspired this landscape where the rationalist perspective, the monumental buildings, spectacular lights, gloomy tunnels and sport reminiscences permeate the atmosphere of the model.
V.Cagliani + M.Grañena + L.Verano

05-NUS DE LA TRINITAT
INNER MOVEMENT

As static and dynamic defines the structure of the Trinitat Nus Park in the previous model, here the terms are reversed and the park activity is associated to movement, as a thrombus, while in the part where there is no activity is static.
Eva Rosenova.

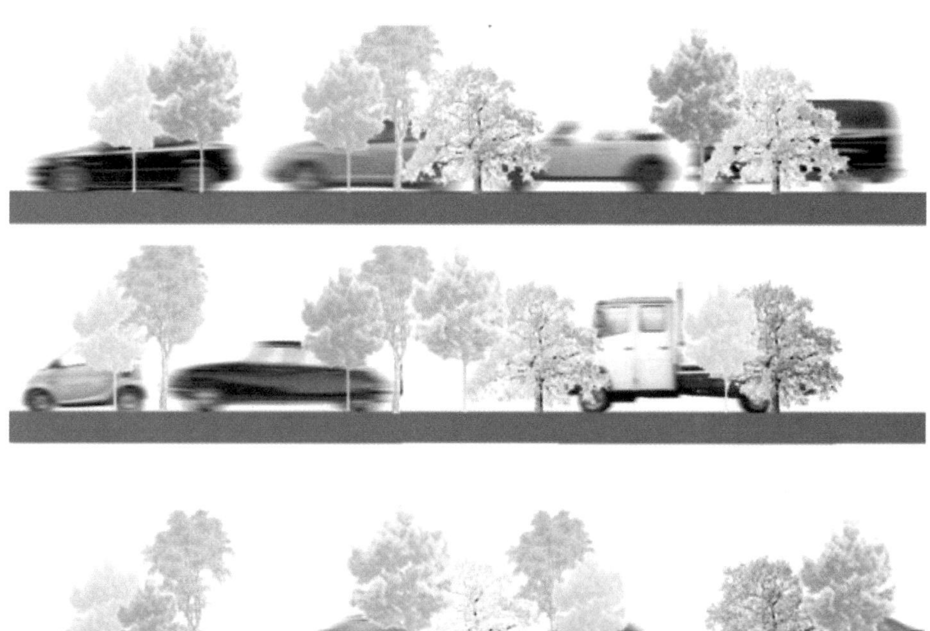

5. THE URBAN INFRASTRUCTURES AS A FORM TO REMAKE THE CITY. Miquel Vidal.

The interstitialities, the in-between spaces mentioned in the villages of the plain of Barcelona, experienced a major occupation with the creation of the second belt of Barcelona. It was designed and built in the context of the Olympics with its philosophy based on the "Barcelona model" to consider the intervention in a public space as a reference for a new model for the city. The Park del Nus Trinitat by Enric Batlle and Joan Roig, is a city park of 7,400 m^2 inserted in one of the most important circulation knots in Europe. The park adopts a circular form immersed in the clothoid circulatory system, which organizes its entire design. Equipment and dunes are placed in the central circular space with a surface of water between them in order to protect the park from noise.

This work refers to duality in basic concepts at the Park del Nus de la Trinitat. This duality is observed between movement-pause, the human scale-the territorial scale and open-closed spaces.

05-NUS DE LA TRINITAT
STATIC VERSUS DYNAMIC

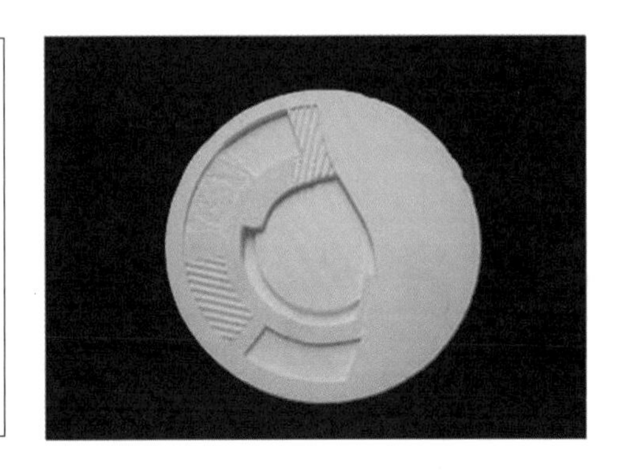

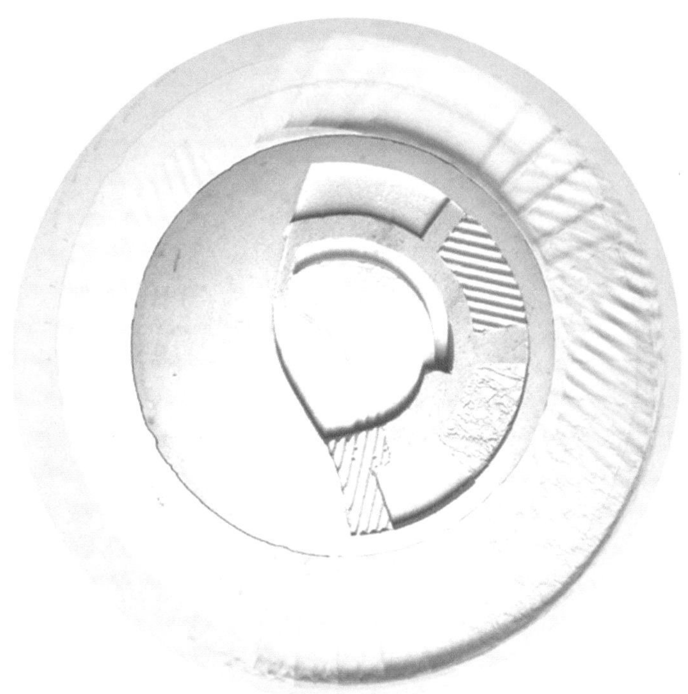

This model seeks to highlight the static part, which refers to the park's surrounding comprised of a dynamic tangle of highways, links, inputs and outputs. The model is heavy and round, static. The functions are explained giving a gentle texturing to the skin of the model.
Johan Backman.

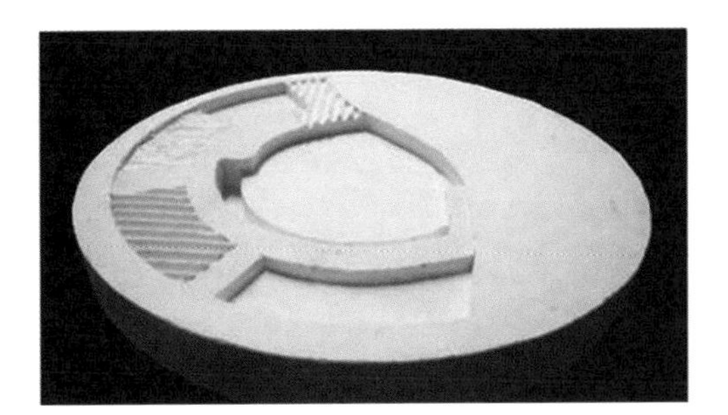

06-BESOS *MASSIVE ORIGAMI*

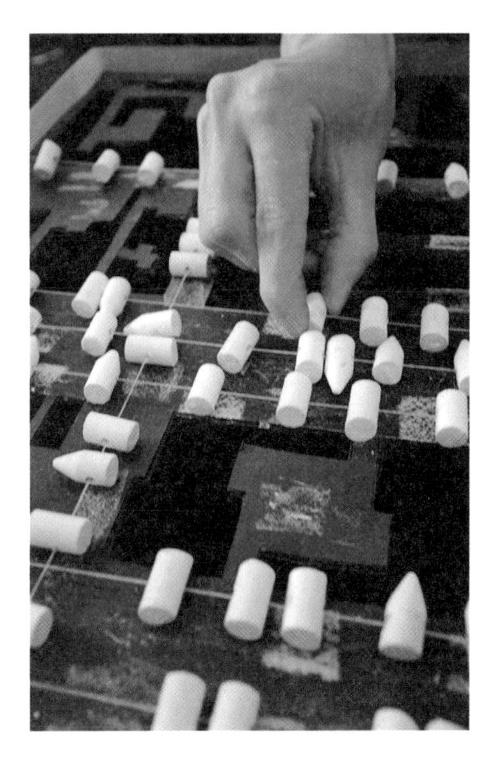

Noticeable units along a main axis organise Besos quarter. The urban grid is completed by transversal, less relevant, streets. This composition, with Forum in its end point does not cover up the repetitive and massive character typical of 70's urbanism.
The masterplan was meant to meet the rising housing demand while searching for new dwelling standards far from the old packed city layout. These apartment blocks are organized around central, strategical facilities and green lots, shaping the core of pieces of a larger system. The base of the model is built as a negative image, acts, also, as the heavy urban grid, highlighting the open spaces and its profusion.
Moreover, these open areas without real identity, due to cars presence are turned into parking lots. As a result, a new layer is added in the model, a moving stratum, symbolizing the dynamics, the traces, lines draw by cars as the leading tool of Besos' public areas and its subsequent deterioration. In short, the model suggests a critical vision of massive urban development and the management of open public space.

Paula Anasagasti+Sergi Romero

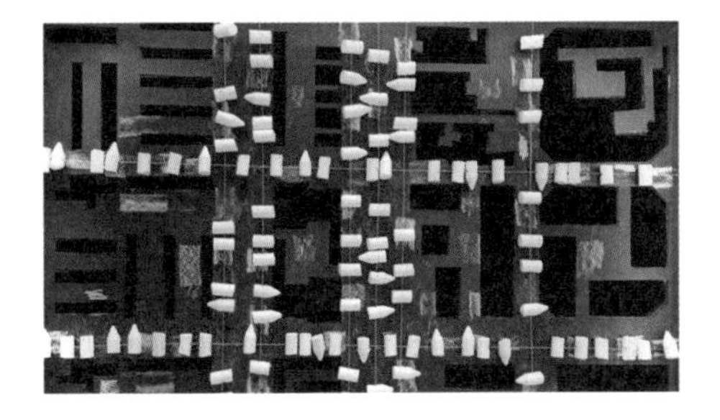

THE POLIGONS OF NEW CREATION. AN ALTERNATIVE READING OF THE CERDA PLAN.
Miquel Vidal.

The district of Besos, originally "the South-West Besos Village", was created on the 23rd of September, 1959 and assumed a different interpretation of the historical distribution of the block Cerdà. It mainly applied to the freely accesible non-built space and the closed interior blocks in the historic Eixample. The poetic model denounces the "contamination" "pollution" of public spaces due to the disorganized invasion of cars.

The great growths of Barcelona colonized the territory from the medieval city and adjusted as between two walls: the sea and the mountains.
But the sea, sometimes longer and becomes a barrier in supporting a fishing neighborhood or at the door to the world of transatlantic trade. Sometimes the city released the urban tumult and offered as an escape in times of political tension. Further sporting, cultural and artistic seize the coastal area for urban development and established there shopping and leisure activities. The latter approach has been sea Barcelona Forum 2004, opening the diagonal.

For a long time the sea was back in Barcelona, like other Mediterranean coastal cities growth came back to the sea turning the waterfront into a place of urban waste. For the 1992 Olympics, Barcelona began a hasty regeneration area between the two towers to the B or s, transforming the waterfront into a public space opportunities for the city. Throughout this journey of the sea approaches, we detected eight different scenarios or interfaces to which man is put between the sea and the city airport city, old port, fishing port, marina, waterfront park, beach leisure , the forum and the technology center B these. The model is full of sensations. In the background the sea. A short recorded sounds from around the coast of Barcelona and this filing on transparencies are eight scenarios offered by the city to the sea. The route is from the Llobregat Kisses and transparencies are removed progressively from the harbor to the giant photovoltaic panel of Oriol Bohigas. During this journey, the man has tested eight scenarios throughout the centuries. Finally, in the last of all, he is discovered beachfront alone, stripped of all suspicion against the soul and exposed as a mine of great value in the open. Somehow the opening of the diagonal is the scenario most spiritual of all, the man and the sea inevitably manifest . We have abandoned nets, poles, piles of metal containers, cranes with giant arms, flags, umbrellas, sailboats, sand castles ... there is nothing to interrupt the speech between man and the sea...

The board forum and s one of those places that reveal the nature and extent of our limited condition of human creatures. T he boundary between land and sea often contain mystery, d the same way as the Orkney Island the UK, here are produced each day infinite dialogue impossible with the sea. now is anecdotal plate passive solar or other energy technology pileup. However, the man feels that flows with the strength of the earth, and harmonizes your instincts beauty under the protection of a titanic dolmen which places limits on the scale and the planet.

07-BARCELONA SEA FRONT
THE FORUM

"Being born without a past, without anything prior to refer, and then to see it all, feel it, and must feel the dawn leaves dew receiving, opening the eyes to light smile, bless the morning, the soul, the life received , what a beautiful life! There being nothing or almost nothing why not smile the universe, the day you go, take time as a wonderful gift, a gift from a God who knows our secret, our inanity and does not care, we do not save grudge for not being ...I'm free ... And as for that being, he believed he had, live simply, let go the image I had of myself, as corresponds to nothing and all, any obligation of coming to be me, or wanting to be" .Maria Zambrano.

7. BARCELONA'S SEAFRONT AS AN ACCOUNT OF SUBORDINATION AND OPPORTUNITY.
Miquel Vidal.

The relationship between the city and the sea is complex and variable, consiting of: the commercial port, the door and urban front to leisure and cultural space, and the industrial space or obsolete Olympic space.

Transformation and experiences are expressed in the model via an accumulation of images projected inside a dark box. The images are shown in order to express the dream of the city.

北京部分

01-NANLUOGUXIANG

BALANCE: BETWEEN THE TRADITIONAL AND THE MODERN

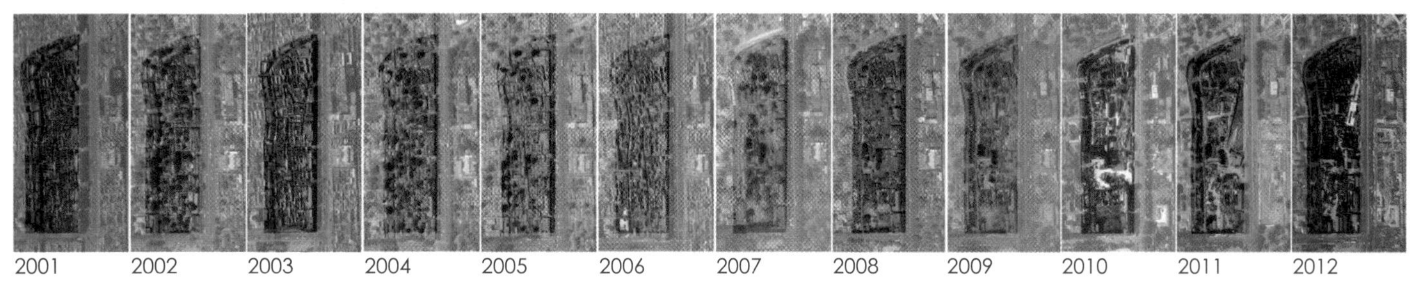

Beijing, the capital of China, has a history of more than 3,000 years since the city is built. Just like other cities in the world, with the development of technology and the change of life, the original form of the city is unable to continue to meet the functions of the modern city. Beijing has accelerated the development of urbanization process. With the changes of the structure in the old city, the original public life disappeared, intimate neighborhood relations have gradually been replaced by the fast-paced lifestyle.

Within less than a decade, many areas of the Beijing city has been changed. The streets and buildings which were maintained for centuries has already been demolished, a new mechanism and structure is developed within this transformation. A more convenient transportation and a more efficient way of life replaced the original slow-paced comfortable lifestyle. The foundation for the further development of Beijing city has already been carry out.

The past stories of the streets, the old way of life, the leisurely elderly under the dancing shade of huge trees and the close neighbors has being remembered. Besides, people is also like a modern mode of transportation, convenient shopping and travel, better community services and new apartment. For Beijing which has a long history, how to balance the relationship between tradition and modernity is the key to the development of the city.

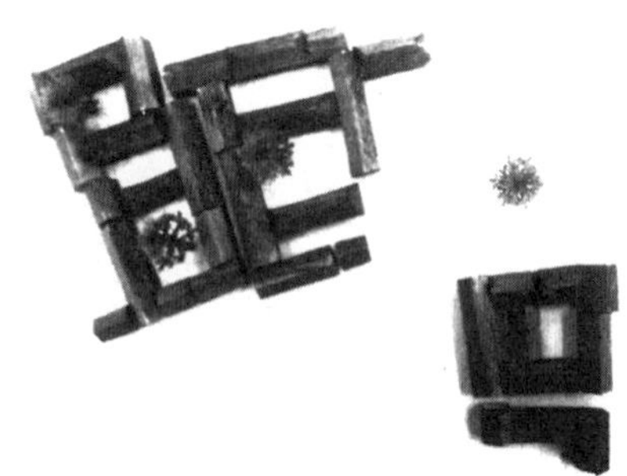

The Nanluoguxiang, which is a very well-known historic district of Beijing sets a typical example of the contradict process of the balance. In the past decades, the old houses have been demolished or converted. The new building has to be intervened into the texture of the old city, and fundamentally change the morphology of the urban areas. The model used in the exhibition intercepted of a fragment of the old blocks to show the changing process.
You can use your own hands to accomplish the changes in this area and understand the process. You can also try to reorganize the form of this region, so as to achiveve a balance of both traditional and modern.

02-HAIDIAN SLANTING STREET *MEMORY*

Haidian slanting street is located in the western suburbs of Beijing.The Haidian town initially rised In the avenue leading from ancient Thistle City to Juyongguan. The direction of the road which contacted from the capital to Haidian was changed from north-south to northwest -southeast. It appeared a avenue to the western suburbs. During the Ming and Qing Dynasties, the slanting street came into being wihth the extension of Haidian block. During the Qing, with the rise of the imperial gardens in the area, the slanting street became the royal road leading from the Forbidden City to the royal gardens in western suburbs. Paved with the granite stone, the road was in high specification, it was a few meters wide and the gutters were on both side. It was an important identity of street layout in the ancient town of Haidian.

Haidian slanting street tells the story of the Zhongguancun and Haidian town, form a station on the royal road to the high technology center. In order to represent this historical and social process, the model has four parts: royal street, village, electronic market and fashion shopping pedestrian street. At present, Haidian slanting street and the village have no longer existed, only a little part of the road was preserved. Watching the bustling city, how to deal with the relationship between site history and city development is a hard problem that we need to continuous explore.

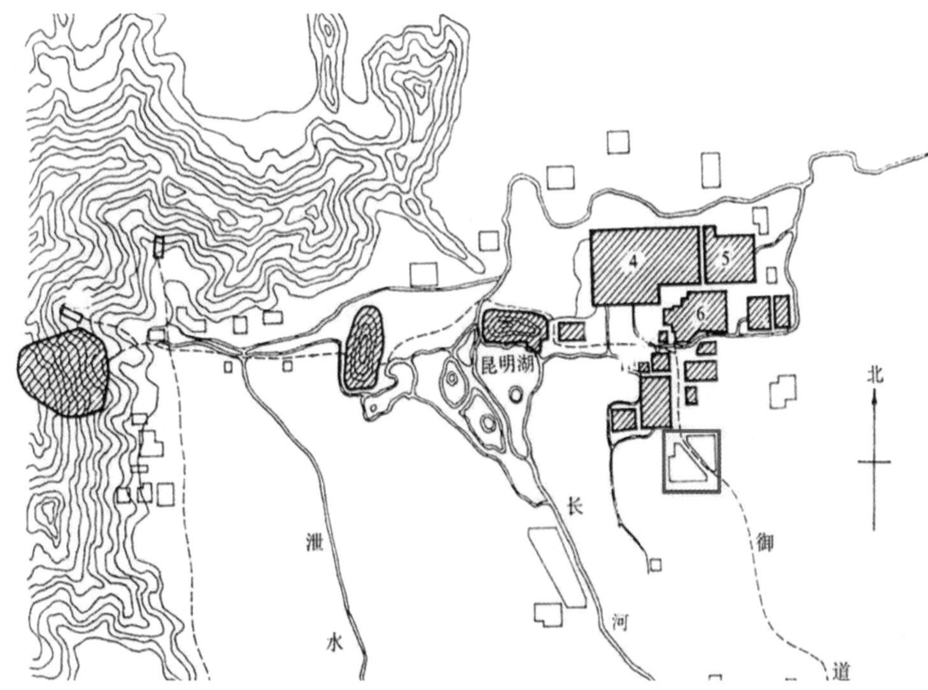

03-798 ART ZONE CONTENT

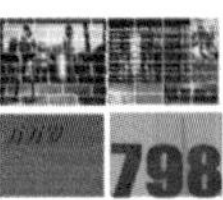

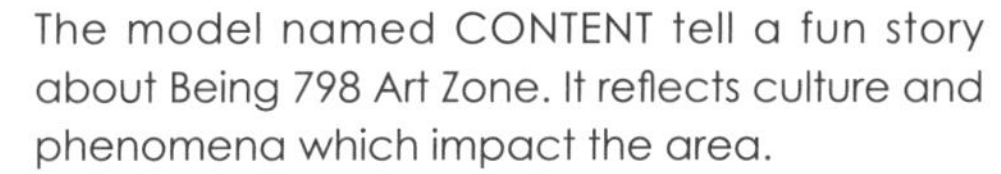

The model named CONTENT tell a fun story about Being 798 Art Zone. It reflects culture and phenomena which impact the area.

The model CONTENT is similar with the book CONTENT, which is wrote by Rem Koolhaas, the Netherlans famous architect. The book CONTENT has very rich content in it, just like a magazine. The book reflect the changing geopolitical relations since 9.11, architecture and politics, history, science and technology, and society. The model CONTENT just to say what has happening in this area. It also could be played as a game.

Attracted by ordered designing, convenient traffic, unique style of Bauhaus architecture, many art organizations and artists came to rent the vacant plants and transformed them, gradually formed a district gathered galleries, art studios, cultural companies, fashion shops etc. As the earliest area where the art organizations and artists moved in located in the original area of 798 factory, this place was named as Beijing 798 Art Zone.

Not only can 798 Art Center hold diversified art exhibis, but also provide space for commercial activities. Many international brands such as Bens, Dior, Huge Boss, Nike, Mark Cheung etc. have held promotions or press conferences here, which make their product possess creative value. Moreover, 798 Art Center with the functions of information distribution center of 798 Art Zone can supply the latest exhibition information and the art history of 798 Art Zone for the visitors.

REPUBLIC
futures existence interest color thought-provoking independence Curiosity speculate
light of happiness possibilities Wisdom commen sense nature
Speak with the future International Cooperation
interactive playground Culture Program dream Art Program
Public Program Press Collaboration Culture collaborative social services
international Research Children/Youth Program Art Cinema
Read Freedom muse Knowledge Live Performance
Power imagination embrace infinite
Beauty free entangled minds

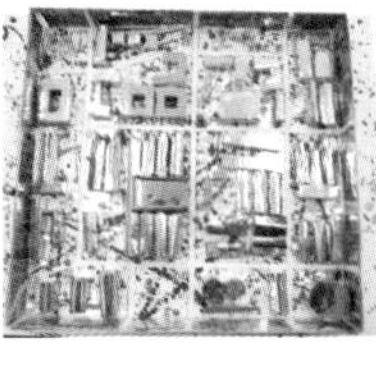

Mottled red-brick wall, Scattered orderly industrial plants, crisscross pipelines, slogans of different ages on the wall. Uniformed workers and fashion visitors form a unique scenery. History and reality, industry and the arts perfectly fit here.

Beijing 798 Art Zone lies in the 718 factory of Beijing Sevenstar Science and Technology Co. LTD.(short for Sevenstar Grop) . A total construction area of 230,000 square meters assembles many cultural elements such as galleries, design studios, art exhibition spaces, artists' studios, fashion shops, restaurants etc. In 798 Art Zone, there are nearly 400 organizations including galleries, artists' private studios, cultural companies like animated cartoon, television media, publishing, design and consultation.

Beijing 798 creative art festival: hold from the end of September to the end of October, focusing on exhibition and communication of culture and art.

In the future, the 798 office will attract lots of famous artists and art organizations from domestic and abroad, according to the instruction policy of “protection, exploration, stability, development” , to build Beijing 798 Art Zone as the center of the exhibition and trade of culture and art, as the most unique and influential cultural creative industry basement in Beijing and the worldwide cultural creative industry park.

04-CCTV
FOLDED BUILDING

Beijing is a famous oriental city with profound history and splendid culture. In this city, the clusters composed with courtyard house, tenement and "hutong" formed, the basic units of the structure of urban space. Today, the traditional structure of urban space is being folded up with the rising of new buildings. As the new urban planning broke traditional living space, the people lost their old life style. However, the loss of the change outweighs the gain sometimes. As the structure of the city is changing, certain spaces are given birth. The newly developed urban space has the power to reshape the identity of the city. How to use them? Put it into building development or green space? Our decision will influence the future of Beijing.

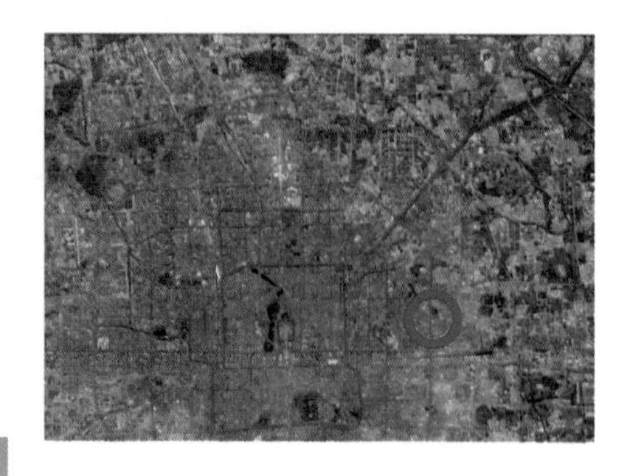

The old map of Qing Dynasty, which is on the base, represents the traditional culture of the city and the grid pattern system. The new building, CCTV Tower, which is made with paper, is the symbol representing the new buildings that change urban space and landscape. The new building has devastating impact to the original city. The effect of the devastating impact is over the building sites, and influencing a wider scope.

05-AIRWAY
THE GAME NEVER STOPS

Weiqi, a game played with black and white pieces on a board of 361 crosses. Any piece is live by airway, the airway come from crosses. The pieces of a single piece on the board, with its straight lines next to the empty points are the pieces of "Airway".For example, all of the airway is taken up for each other, it is airway – free status. Both the black and white ones are living on the airway in the middle.

How to deal with old and new architecture and heritage, to build the harmonious relationship between new and old landmark is a focus issue challenging the ancient capital of Beijing urban metabolism.

The process of urbanization in city is a game of chess, a development of the metabolic process. To keep the metabolism you have to aerobic respiration breathe the mouth. So we can join in some airway between the old and new buildings to complete harmony and make balance. The "airway " can be open spaces, ecological corridor or green square during the dense construction.

What we should do is maintaining dynamic balancing by the airway.

On endless Game in the urban development.

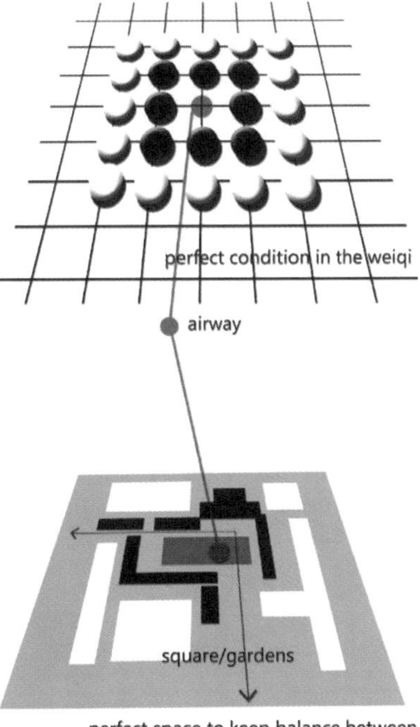

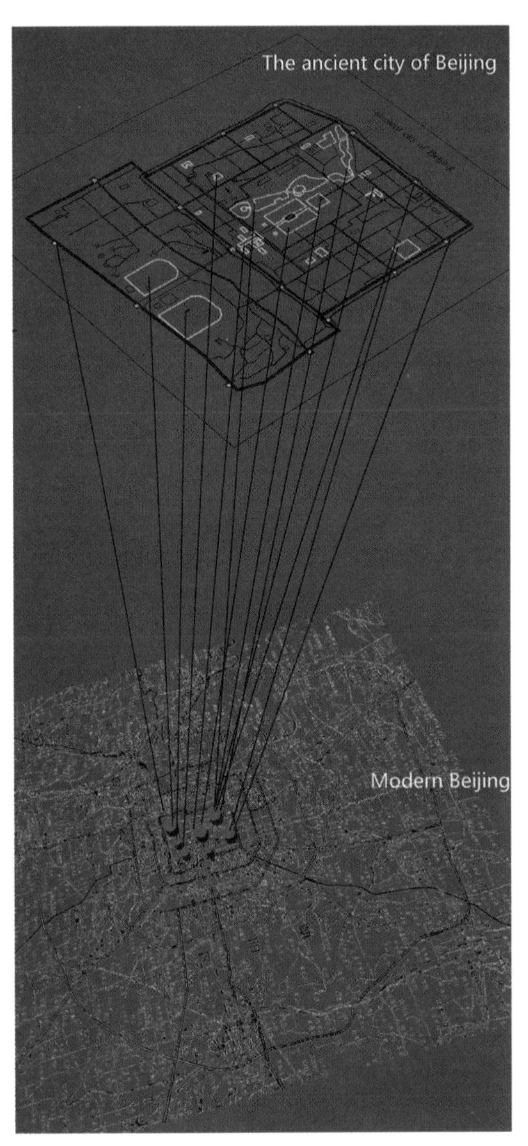

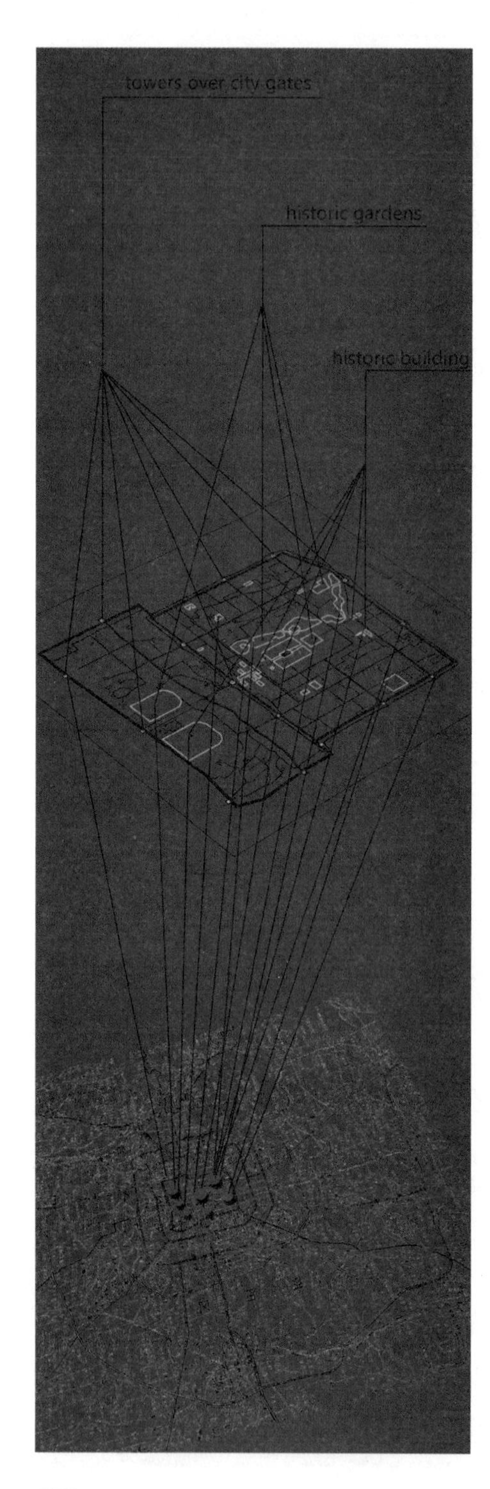

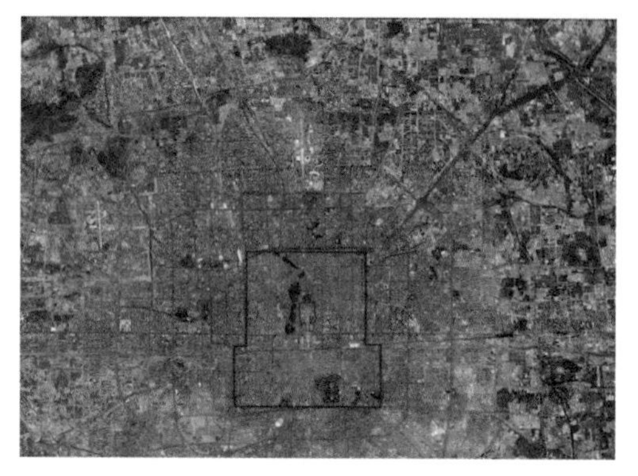

Black Stone is the classical urban heritage of building full of cultural value.
Part in the air, represent the ancient buildings have disappeared. In the part of the floor on behalf of ancient buildings still intact.
Representatives of the white stones is simple, modern, multifunctional buildings.

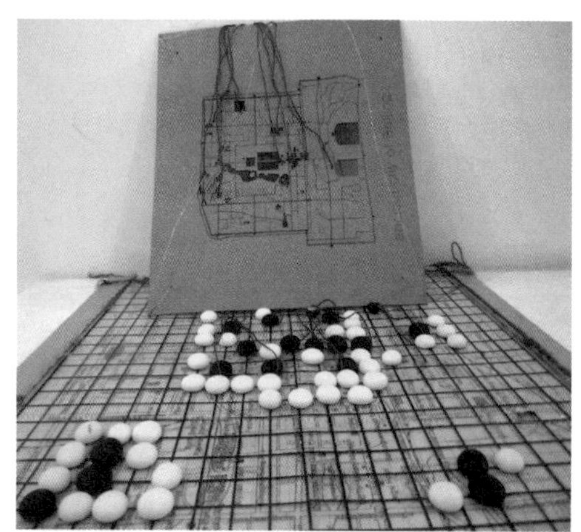

06-COLOUR

The barrel comprises a map of Beijing. In this map, there are a lot of pierced holes. These holes symbolize the green system of Beijing. As time going on, the colorful of the light through transparent film changes every second. It reflects the different seasons, different plants form the color of Beijing. You can rotate the handle then you can find beautiful color.

Beijing is a city with four distinctive seasons, each season has its unique color. The color of the main carrier is the green space which all over every corner in the city. The green system of Beijing has different spatial levels and different types of green space. The green space can provide Beijing with a wonderful landscape in four distinctive seasons, which become a characteristic feature of Beijing. At the same time, the green system of Beijing provides the diversity of space for activities of the citizen. The city green land becomes the highlights in whole city.

07-798 ARTS ZONE
SPACE-TIME DIALOGUE

The model shows the existence and development of the 798 as the most culturally iconic city arts center of Beijing, with iconic architectural language and abstract art messages, it also shows the capacity and future potential of Beijing as the capital of the world. Refined the serrated plant belonging to the style of the typical German Bauhaus art, and transformed the three buildings to art sculpture, its interior modern art can be observed through the side of the wall scattered window, and the other side of the wall reflects history with simple and pure manner. Three buildings tell of the historical changes of the early days to the present, the smallest one is about "slogens" in New Era of China, the middle one is about "the electronics industry" in the period of 718 Joint Factory, and the biggest one is about "arts" in the new period of the Beijing City Art Center. These are a true reflection of the plant at different times, to understand the modern colorful world.

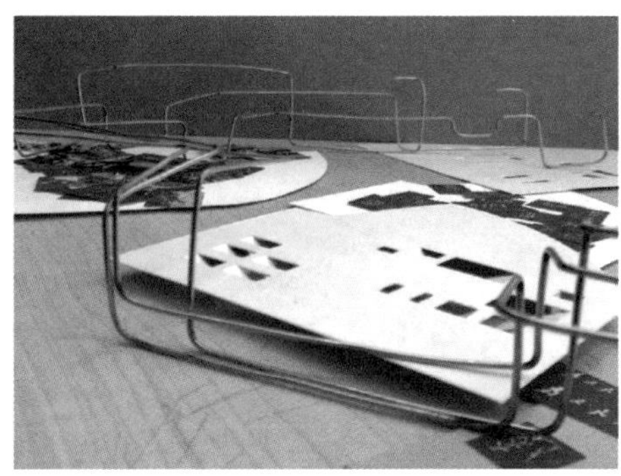

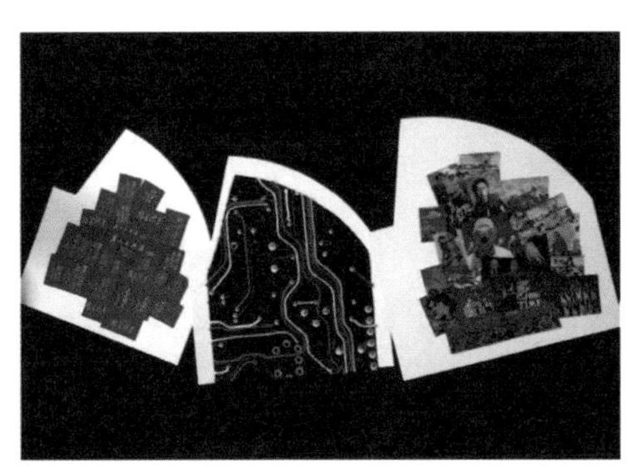

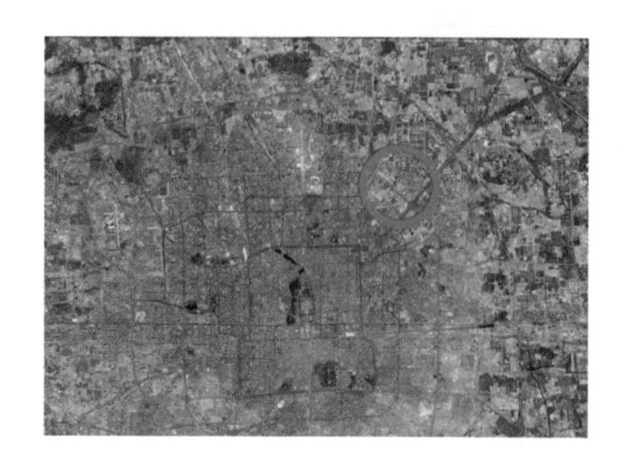

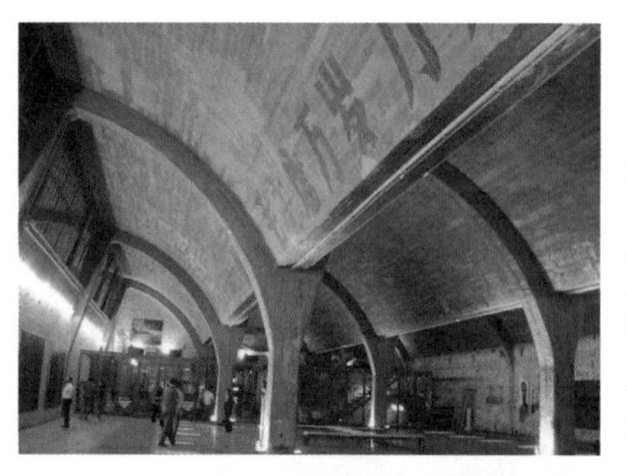

798 Art Zone, most culturally iconic City Art Center.

In 1952, the joint factory to be built in Jiuxianqiao region without any industrial base in the surburbs of Beijing, civil constrained began in 1954 and the factory started production in 1957. Planning to build 774 Factory, 738 joint factory have made outstanding contributions on the country's economic construction, especially the construction of the electronics industry, national defense construction, the development of the communication industry.
718 Joint Factory was one of the assistance projects in China from the help of the socialist camp. As the electronics industry's leading position in this camp, East Germany was given the impossible task of building the joint factory. The joint plant designed by the German, typical Bauhaus style. Thie whole plant planned orderly, with the unique architectural style, attracted many art organizations and artists coming to rent the vacant plants and transform the plants, gradually formed a set of galleries, art studios, cultural companies, fashion shops, and changed into one of the earlist art organizations and artists stationed was located in the original 798 factouy, here was named as Beijing 798 Art Zone, it made an abandoned factory become the fashion community.

08-THE CITY WALL OF BEIJING *WHAT IF?*

60 years ago, Beijing has a fateful choice: one plan is that the center is Tiananmen square. The construction of the administrative center of the capital is on the basis of the anicent Beijing city. Another solution is that open up a new administrative center in the western suburbs of Beijing. History chose the first one, the old city walls which onces standing in the various difficulties and hardships in more than 800 years is destroyed. If we can make choice again, will Beijing be what kind of today?

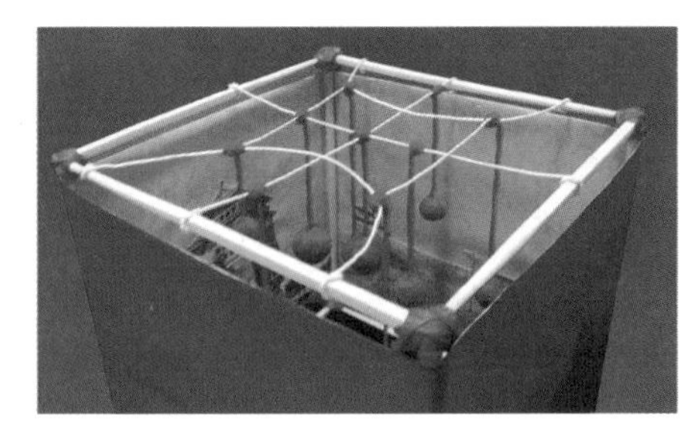

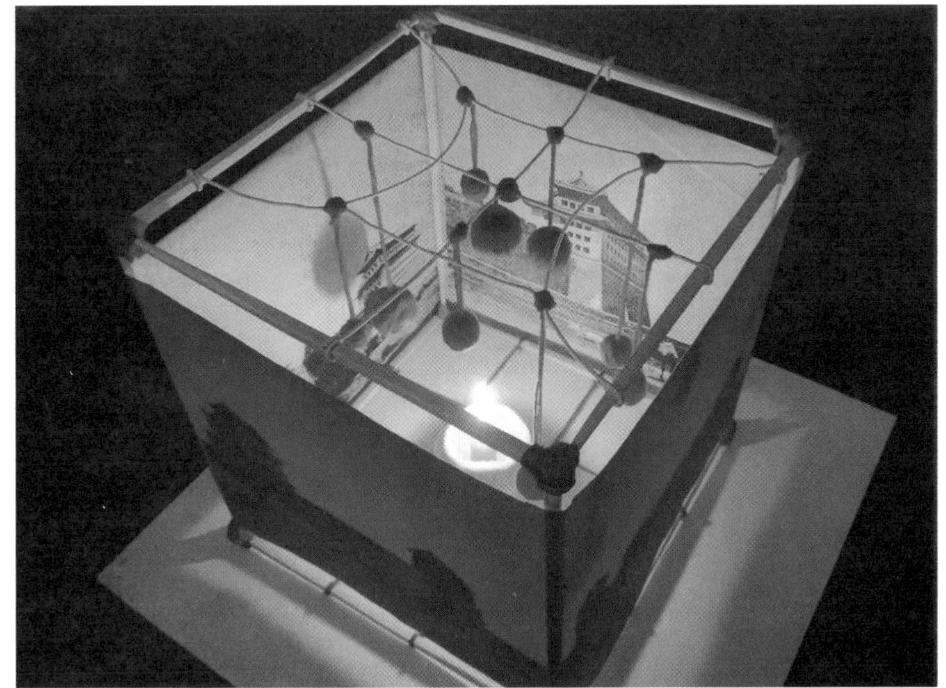

In 1950, Mr. Liang Sicheng and Mr. Chen Zhanxiang put forward a Plan of Beijing City: the location extending the western suburban outside the city is the administrative center of the capital. The West now has a new urban and construct residential area for all levels of administrative officer. The East connects with the old city. The new administrative center, city cultural landscape, the museum district, the celebration assembly square, commercial prosperity area, city administrative district supply equipment, and connect closely the original residential areas in the northern and western part and there is a reasonably short distance so that people can arrive anywhere easily.

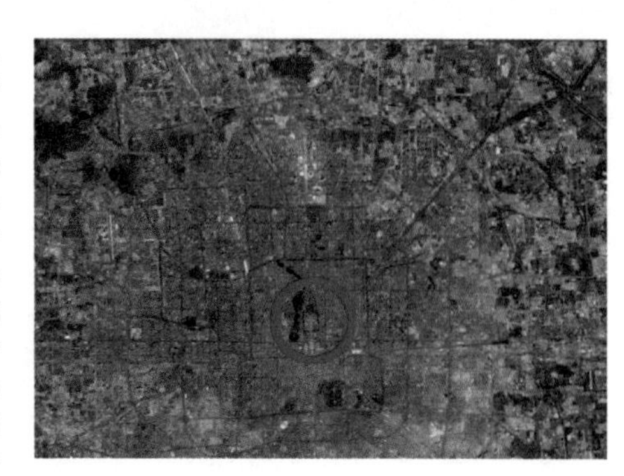

Mr.Liang Sicheng and Mrs Lin Huiyin

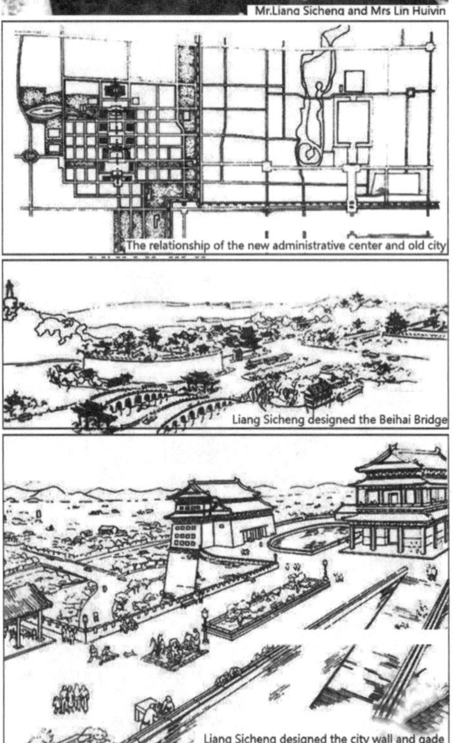

The relationship of the new administrative center and old city

Liang Sicheng designed the Beihai Bridge

Liang Sicheng designed the city wall and gade

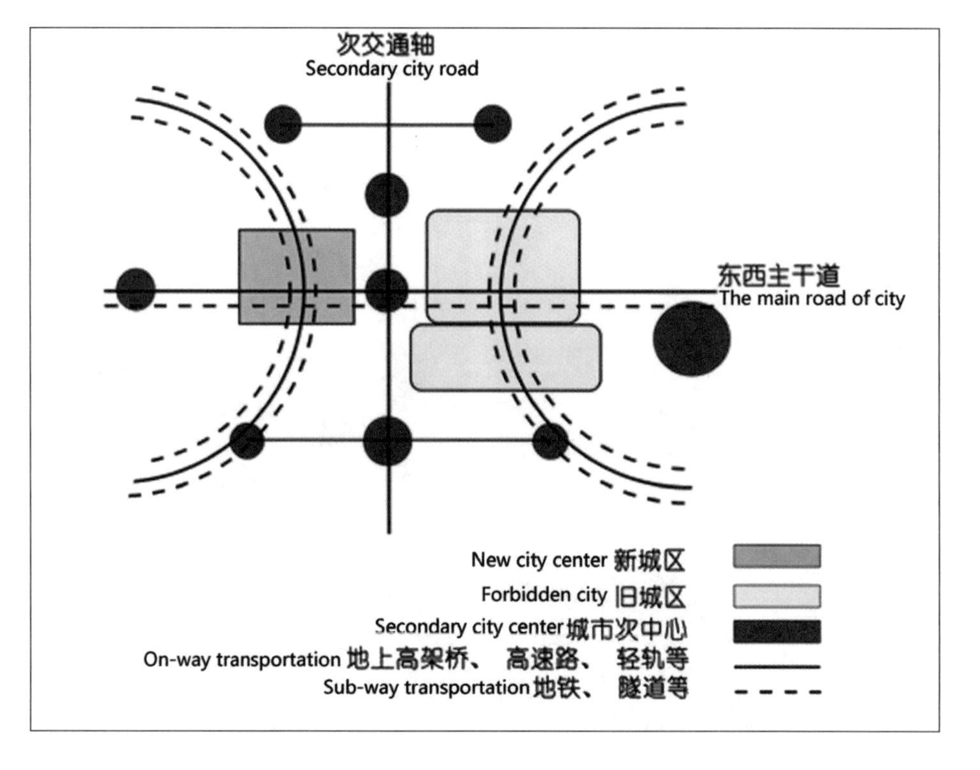

09-THE GREAT TRAFFIC

This model is made of iron wire, small modelcars. And it looks like a circle. The model shows the traffic problem of Beijing. We make the circle go up and down to compare the overpass and the great wall. We guess most people living in Beijing have experienced the jam when they go to work or off work. The problem of traffic shows the limitation of the urban planning, and also shows the richness of Beijing. The westnorth of the model is mush higher to compare the topography of Beijing.

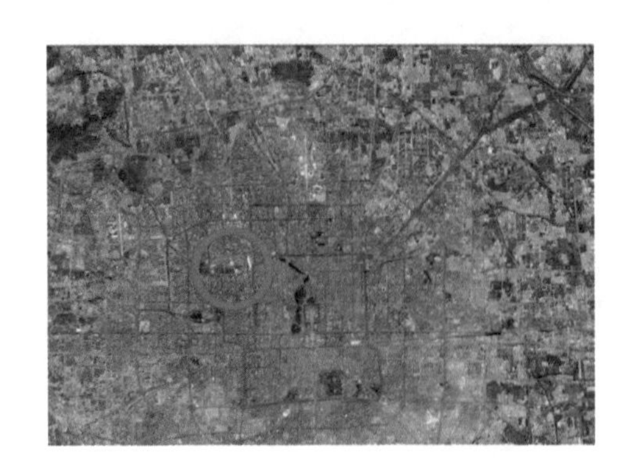

Along with the accelerate progress of the urbanization, the scale of the city becomes much bigger.

More and more overpasses are being built, such as the very famous Xizhimen overpass. 30years ago, Beijing was called the City of Bicycle, but now we might call it the City of motors. Nowadays, cars on road and bought are limited. Such a number of automobiles make air pollution and sound pollution. As the automobiles take place of bikes, people are separated with each other and communicated less. The huge overpasses just look like the Great Wall, and that is the reason we call it the Great traffic.

10-A Bite of Beijing

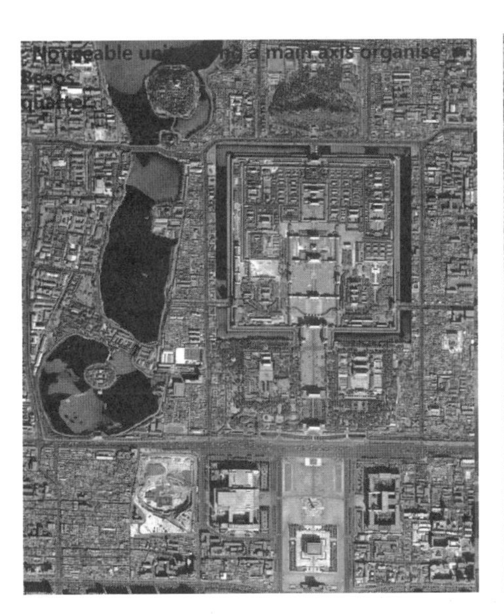

As a international city, the identity of PEKING is diversity. But colorful diet culture which have deeply content showing the screen of PEKING so wonderful.
"Huger breeds discontentment", in one sence the diet stand for the geographical of a city, and the depth information passed from the diet is a concentrated urban regional culture scene. As a comprehensive international city, there are so richful food in PEKING, and more, it is that the various of diet, PEKING achieved the wonderful city view.

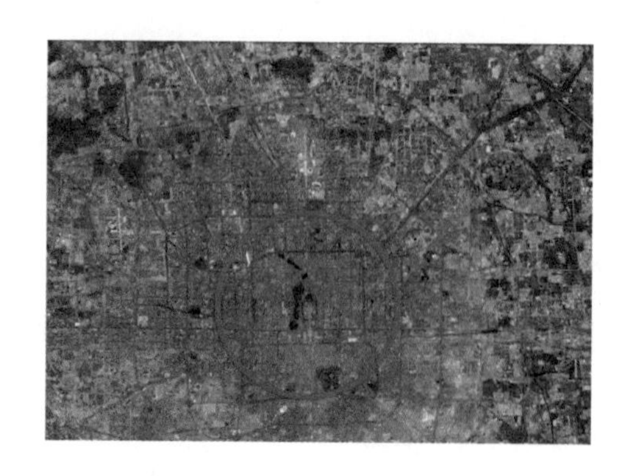

The model use the chopsticks which have connection with diet directly as the primary material to outline the modern Beijing. The combination of chopsticks express the information of the basic physical layout of modern Beijing. In vertically, there are some changing happened to the level of body block, and it is a generalization of the Beijing building outline, while the Horizontal arrangement is a manifestation of the ring urban pattern. Chopsticks as the Chinese people generally use Tableware is used in this model extensivly for tell people that the close relationship between cities and diet. The colorful diet of PEKING also play a importance role in process of Beijing to become an international city.

11-CBD *CHANGE BEING DESTINY*

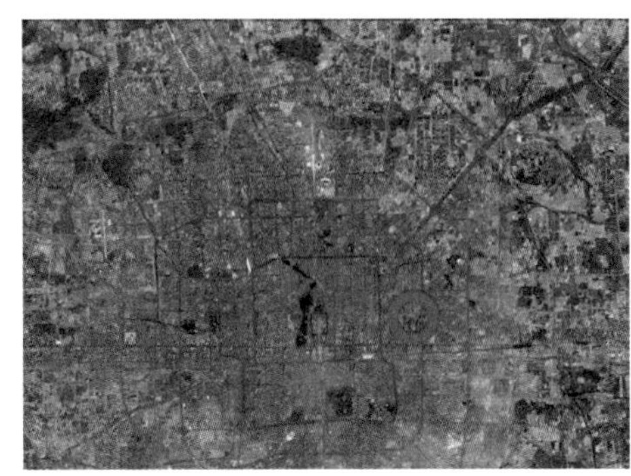

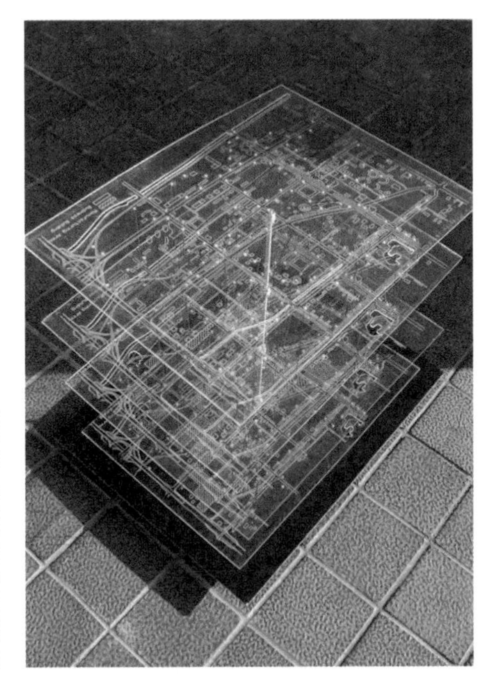

The Chaoyang District was in the east part of Beijing. It attracted many international headquarters companies to setlle in this ares because of the high-speed economic development and the favourable policies in Beijing. Thousands of white-collars created the amazing wealth in the era, then the best shopping center in. So the CBD(Central Business District) was appeared in this area. It's like a city within a city and without a boundary. It symbolized wealth, fashion and culture.

The CBD has changed violently over recent decades. It is an unique experience. In fact, you could identify the site of the CBD when you were in the middle of the Imprial City, because you can see many skyscraper in that direction. They upraised the skyline of the eastern part of Beijing. The limit of the plot ratio and height of the building can be ignored properly and the architectural style and architectural forms can be various, because this area is far from the Beijing ancient capital. The CBD is becoming a "privileged" area of Beijing, and a "building exhibition hall".

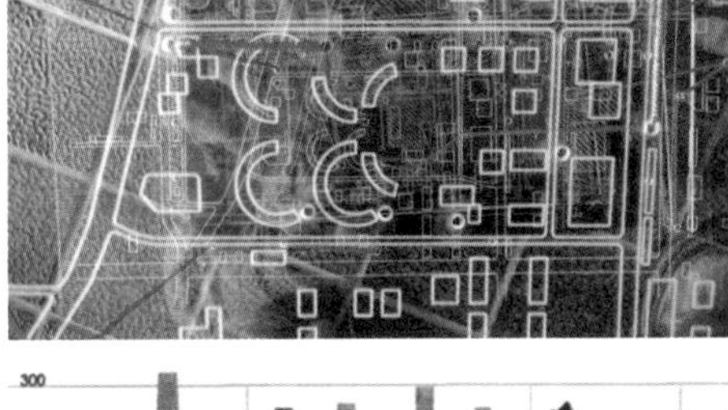

11-CBD
CHANGE BEING DESTINY

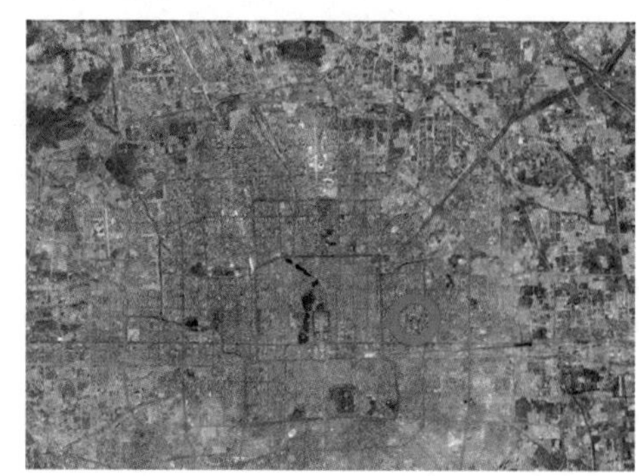

This model shows the changes from 2000 to 2012. This area became the business area from the manufactory disrict.

Although this model shows the changes of ten years, this area has a long history. This area was a royal hunting field in the Ming Dynasty, so it was named the Roe room. In the Qing Dynasty, there are many brick kilns. The largest one locals in the north side. People can see it far away, so it was the mark named DABEIYAO. After 1950, there were many different factories. It became the manufactory district. Beginning in the 1980s, the reform and opening up was starting in China, the DABEIYAO also changed. Because the embassy district was near here, a large number in China began to sttle in this area spontaneously, the East Third Ring Road area became the main area of the foreign office in Beijing gradually.

In 2000, the construction of the Central Business District began.

Now, the CBD will extend to the East Forth Ring Road. It is another change for CBD, and it seems that changing constantly is the destiny of CBD.

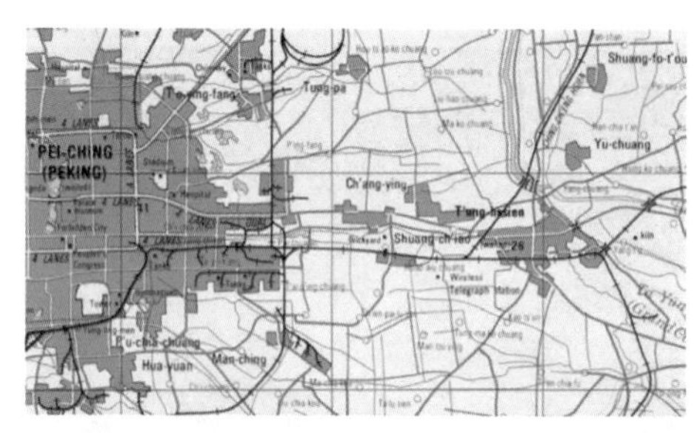

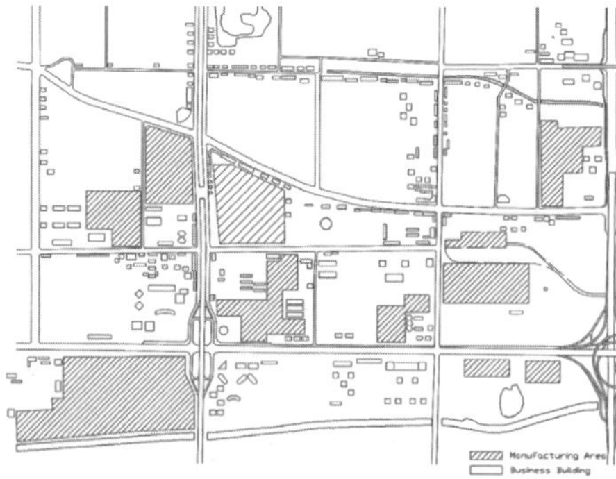

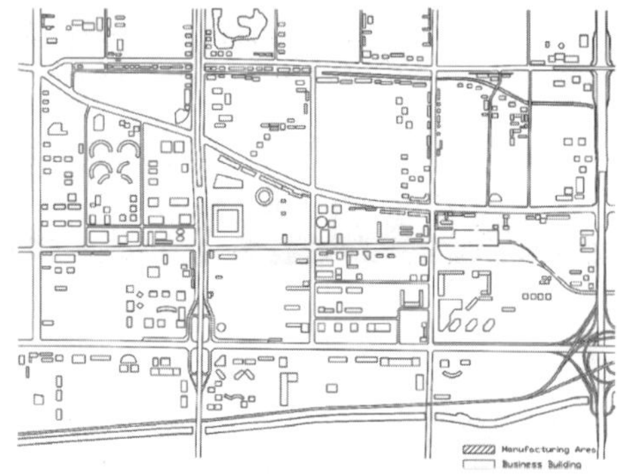

STUDIO PECKING

THU-AA 联合设计课程清华大学部分

2012.9 ~ 2012.12

铺砌城市

Tiling Urban

吕回
Lyu Hui

设计地块位于颐和园西南侧，毗邻京郊三山五园，地势低洼水量充沛，历史上有着优良的自然条件。

随着城市的不断扩展，原本肥沃的农田逐渐被城中村和苗圃取代，生态环境被逐步破坏；场地在未来若干年内会被改建成南水北调工程终点的蓄水池，如何在安置必要生态功能、水利功能，同时妥善解决居民生活问题，成为了设计的要点。

本设计根据现有生态、地质条件为基础，重构地形，安置住宅、商铺、公建、道路、开放空间、湿地、蓄水池等，各项层次空间类型，解决蓄水、净化、栖居、旅游等不同功能。

The site locates by the north-west side of the Summer Palace, where the amount of water was once very rich and many traditional royal gardens were built since 17th century.

However, due to Beijing's urban sprawl, the great ecological circumstance in history was replaced by illegal residence and nursery gardens. Also the government planned to drive off the residents in the site and build the terminal reservoir of the South-to-North Water Diversion Project. The ecological damage, the dwelling of the inhabitants and the Hydraulic engineering became the main issues of this site.

The design generates a parametric mesh as a respond to the issues. In the specific place of the mesh, several prototypes with different functions are installed, such as green residential area, playground, wetland, ponds, etc. The prototypes work together in a systematic way, purifying and storing water, greening open space and motivating tourism.

吕回

Lyu Hui

01 场地现状 Site Analysis

Green System in Beijing

Surroundings of the site

方案位于颐和园外西南侧绿地。设计从场地地形入手，通过水文分析得出适合作为调蓄池的低洼处，并在高处引入住宅、商业等城市功能，通过将建筑组团和景观相互渗透的方式，充分发挥景观的边界效应，一来为当地居民提供家门口的良好生境，二来使得地表径流在进入调蓄池前可以得到充分的净化和过滤，解决雨洪问题的同时保证了生活水源的安全。

Existing Constructions

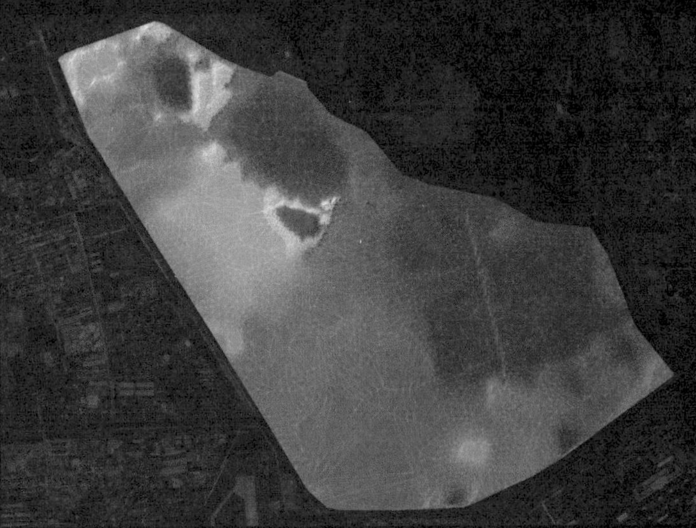
Height of Topography

Tansportation Distance & Density

View Quality to the Yunquan Tower

01 场地现状 Site Analysis

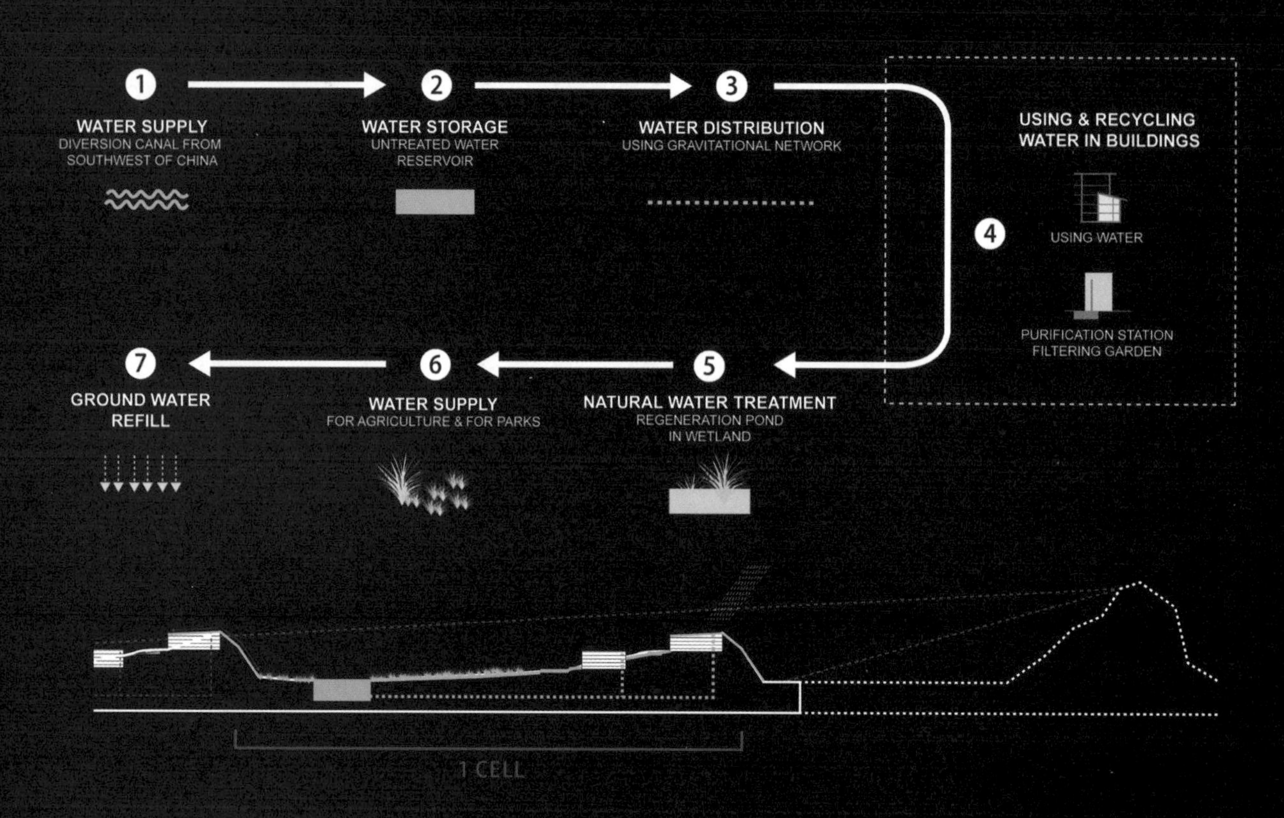

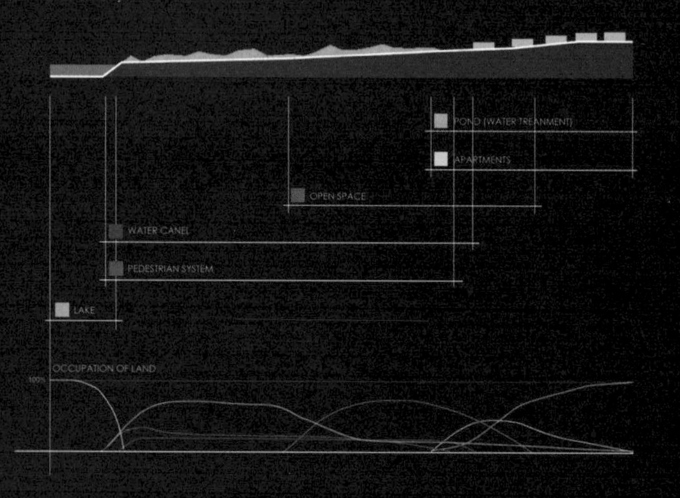

	For one cell	For one tile
Inhabitants	1 500 people	50 people
Housing floor area	60 000 m²	2 000 m²
Number of buildings (4 storeys, 500m² floor area)	30 units	1 unit
Water treatment pond	4 500 m²	150 m²
Water reservoir		-
Re-using grey water in building (25% of total water use)	16 500 m³ / year	550 m³ / year
Green land (agriculture, parks)	22 hectares	0,7 hectares

单元空间结构:

方案根据原有地形重塑地形，利用水流的重力特点安排场地中不同水的使用对象。将水源储藏在低处，利用光伏发电提水给高处的居住区，居住区下水通过设备初步净化后，进入雨水花园进一步净化和下渗，最终清洁的地表径流重新汇入储存池。

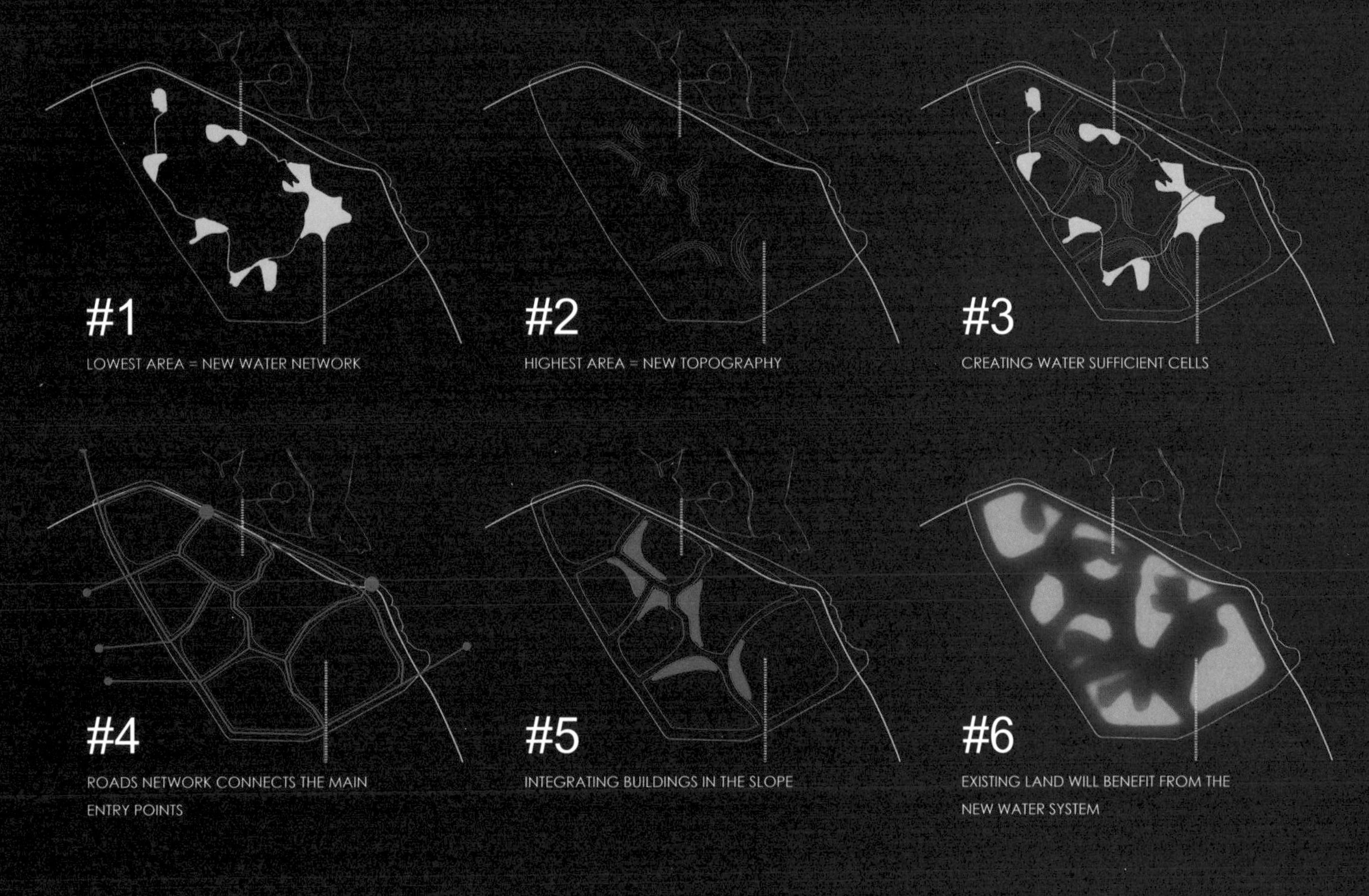

#1
LOWEST AREA = NEW WATER NETWORK
#2
HIGHEST AREA = NEW TOPOGRAPHY
#3
CREATING WATER SUFFICIENT CELLS
#4
ROADS NETWORK CONNECTS THE MAIN ENTRY POINTS
#5
INTEGRATING BUILDINGS IN THE SLOPE
#6
EXISTING LAND WILL BENEFIT FROM THE NEW WATER SYSTEM

设计步骤：

第一步：利用现有地形，在低处开挖若干必要的水利蓄水池，以满足南水北调工程的水利要求。

第二步：蓄水池开挖形成的土方堆积在现状高程较大处，形成若干面向水池的坡面。

第三步：水池和各自周边的坡面形成分别行车各自的汇水单元。

第四步：在每个汇水单元的分水岭上设置连接场地内外的交通网络。

第五步：在道路的两侧，顺着坡面生成台地建筑，形成居住区的相关建筑功能。

第六步：在建筑和水体之间补植地形改造移栽的植物，增加雨水花园的流线，形成建筑到水体间的绿地系统。

01 场地现状 Site Analysis

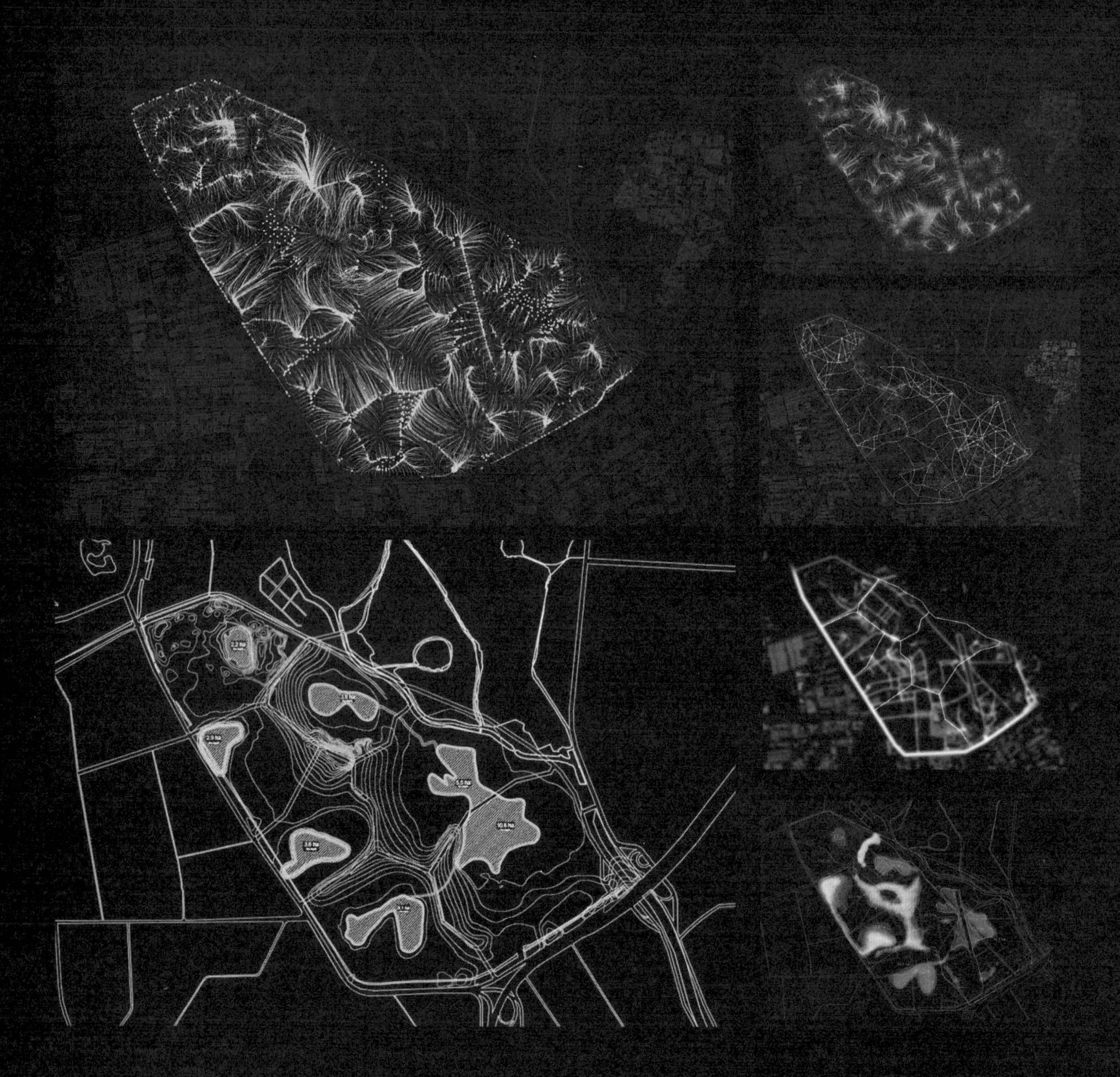

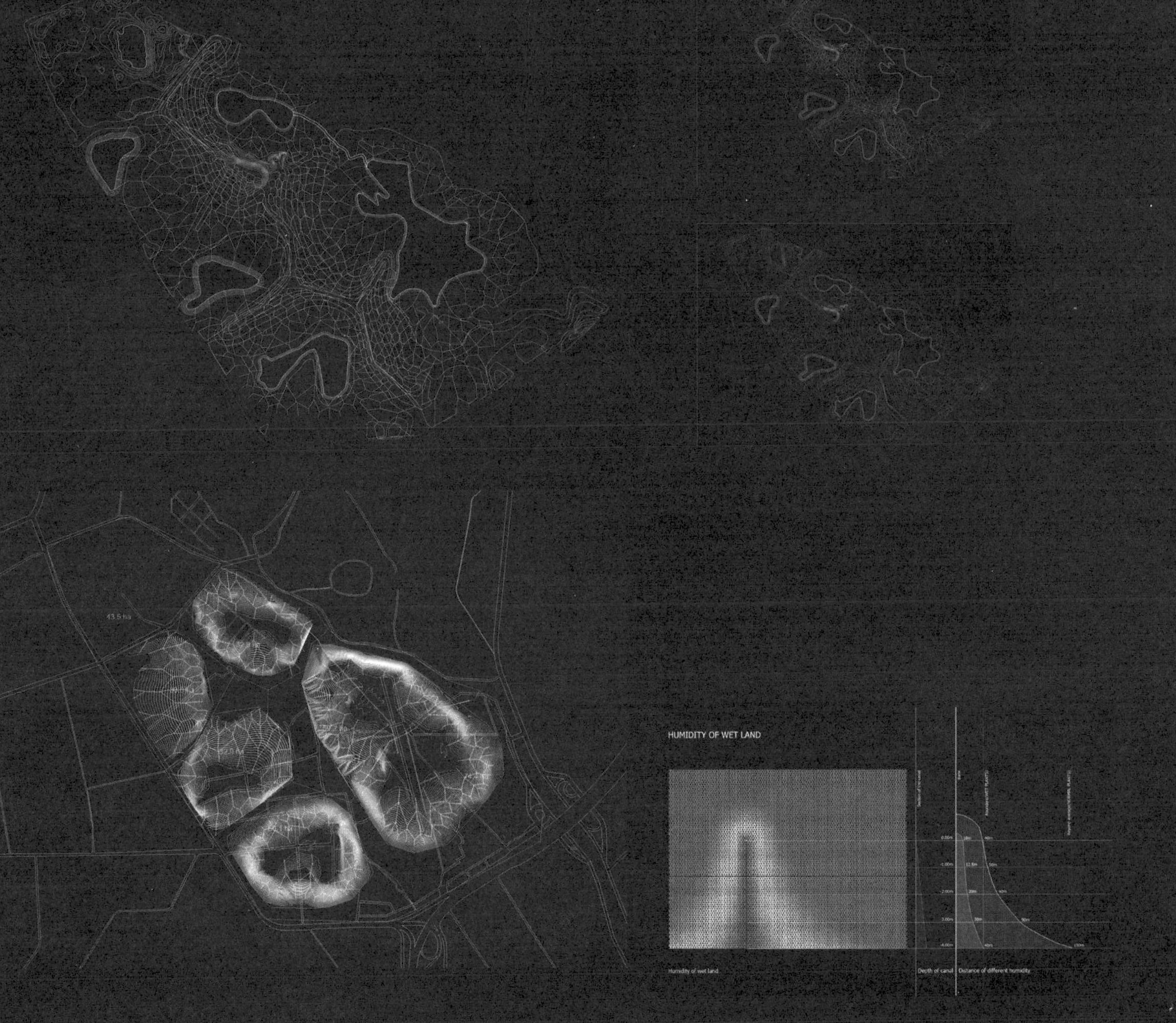
13.5 ha
HUMIDITY OF WET LAND
0.00m
-1.00m
12.8m
-4.00m
Humidity of wet land
Depth of canal
Distance of different humidity

02 理念构思 Concept

1st Floor

2nd Floor

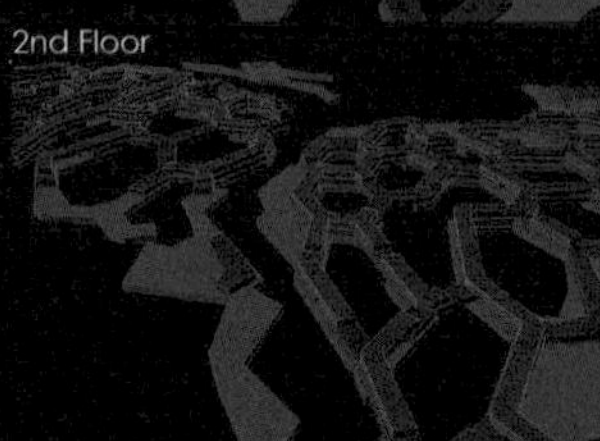

3rd Floor

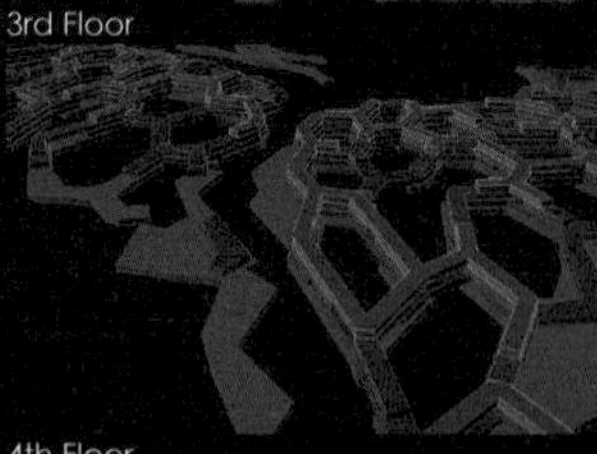

4th Floor

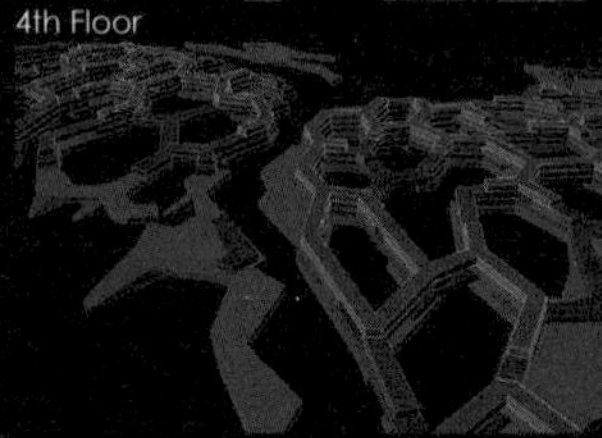

Ramp System

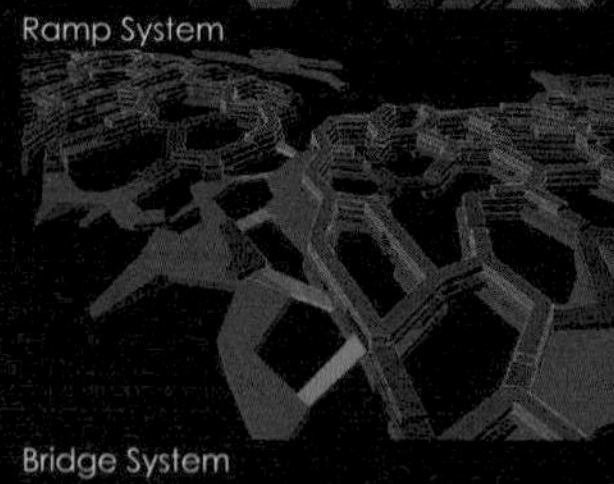

Bridge System

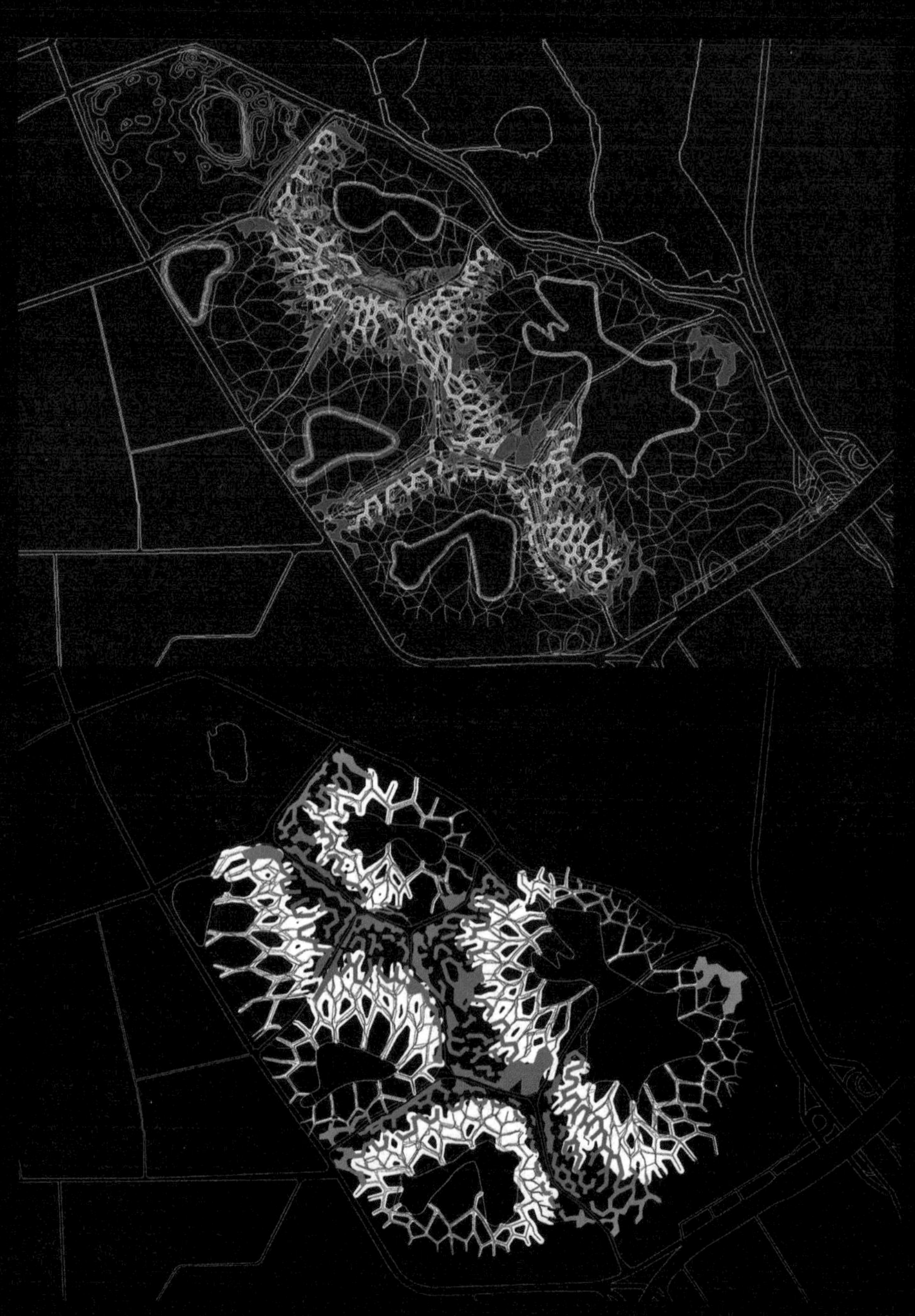

ROW 1,2,3,4 TOP POINTS FIXED

POLYLINE BETWEEN 2 TOP POINTS (2 CONTROL POINTS)

3 CONTROL POINTS

4 CONTROL POINTS

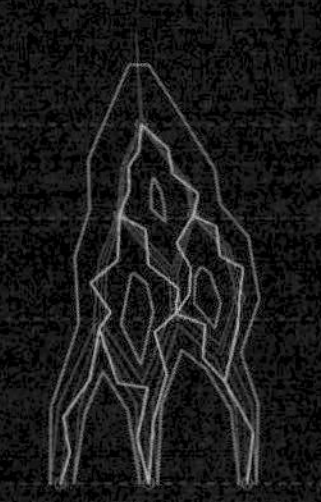

5 CONTROL POINTS

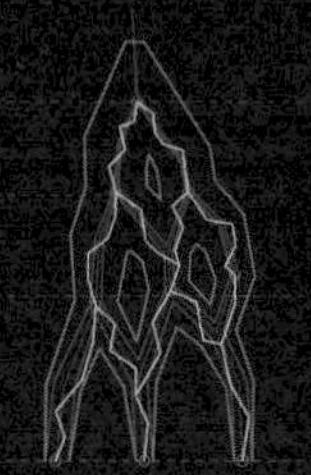

6 CONTROL POINTS

ROW 1,3 TOP POINTS FIXED

POLYLINE BETWEEN 2 TOP POINTS (2 CONTROL POINTS)

3 CONTROL POINTS

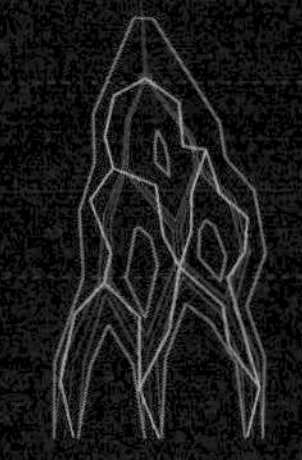

4 CONTROL POINTS

5 CONTROL POINTS

6 CONTROL POINTS

ROW 1 TOP POINTS FIXED

POLYLINE BETWEEN 2 TOP POINTS (2 CONTROL POINTS)

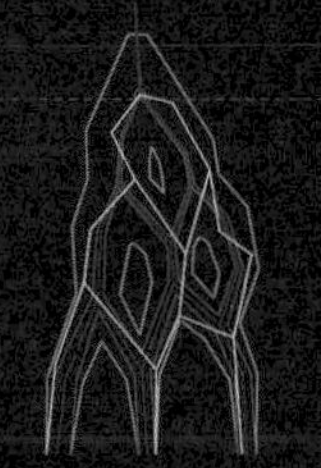

3 CONTROL POINTS

4 CONTROL POINTS

5 CONTROL POINTS

6 CONTROL POINTS

PRIMERY PATH

MEASURE THE LENGTH OF EACH SEGMENT (50, 100M)

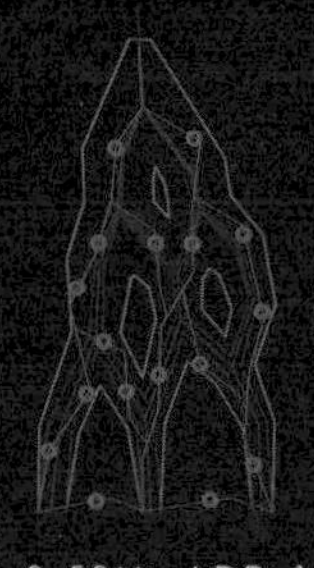

0-50M: ADD 1 ENTRANCE
50-100M: ADD 2 ENTRANCES

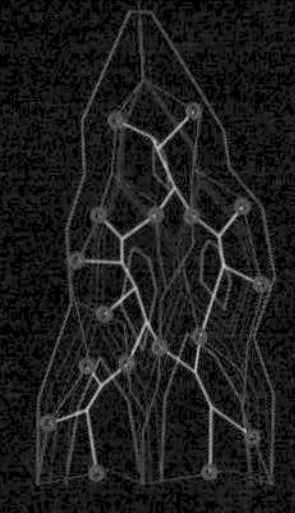

CONNECT EACH ENTRANCE

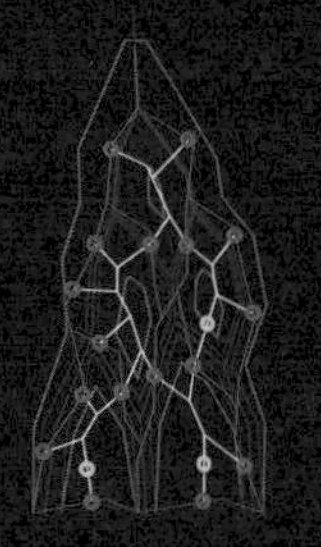

MEASURE THE SECONDORY PATH AND REDIVIED (40M)

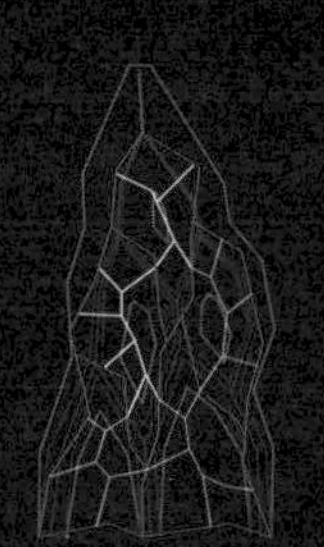

CPMPLETE THE SECONDORY PATH

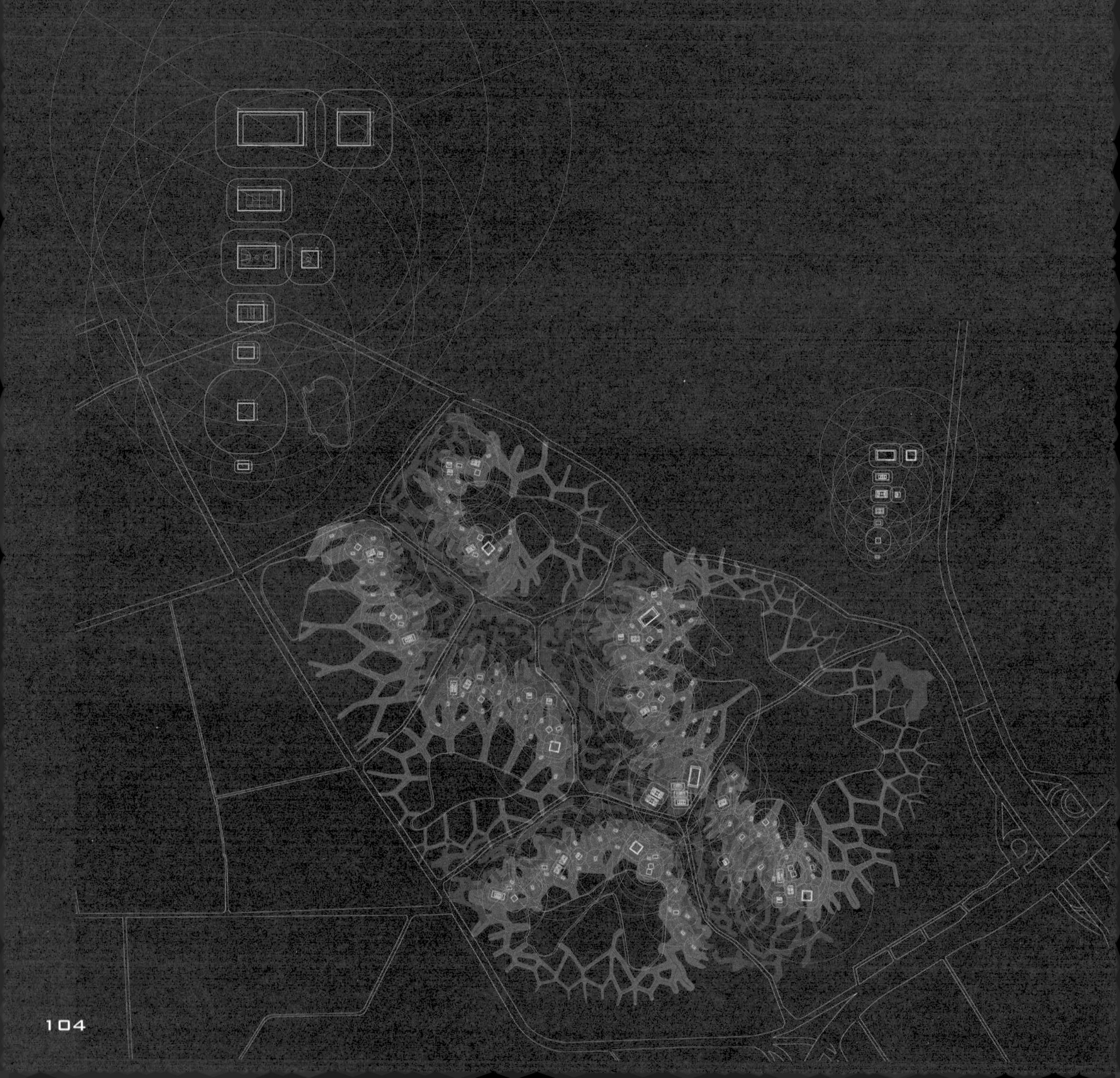

05 模型展示 Model

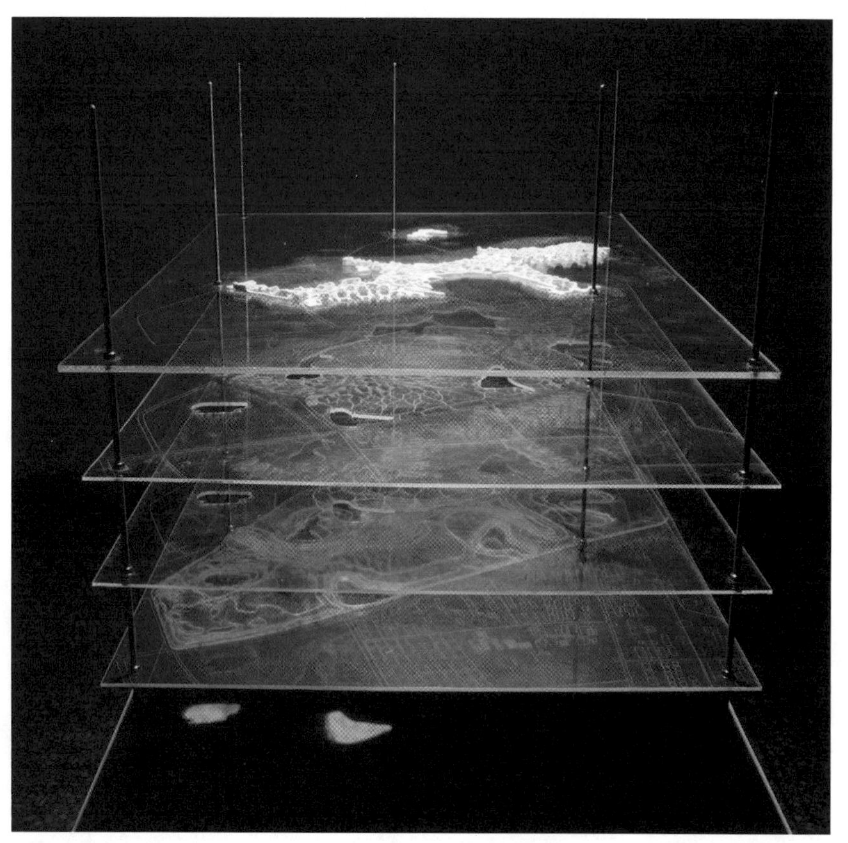

概念模型

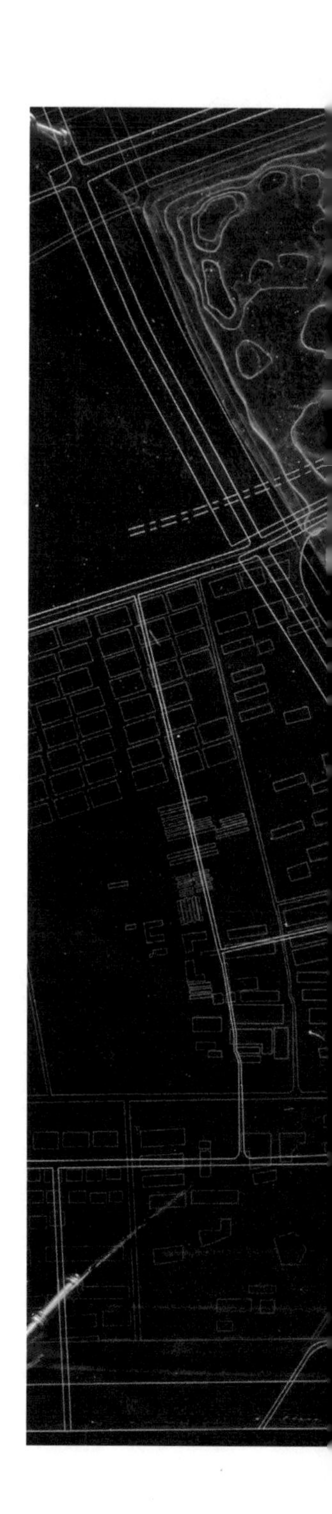

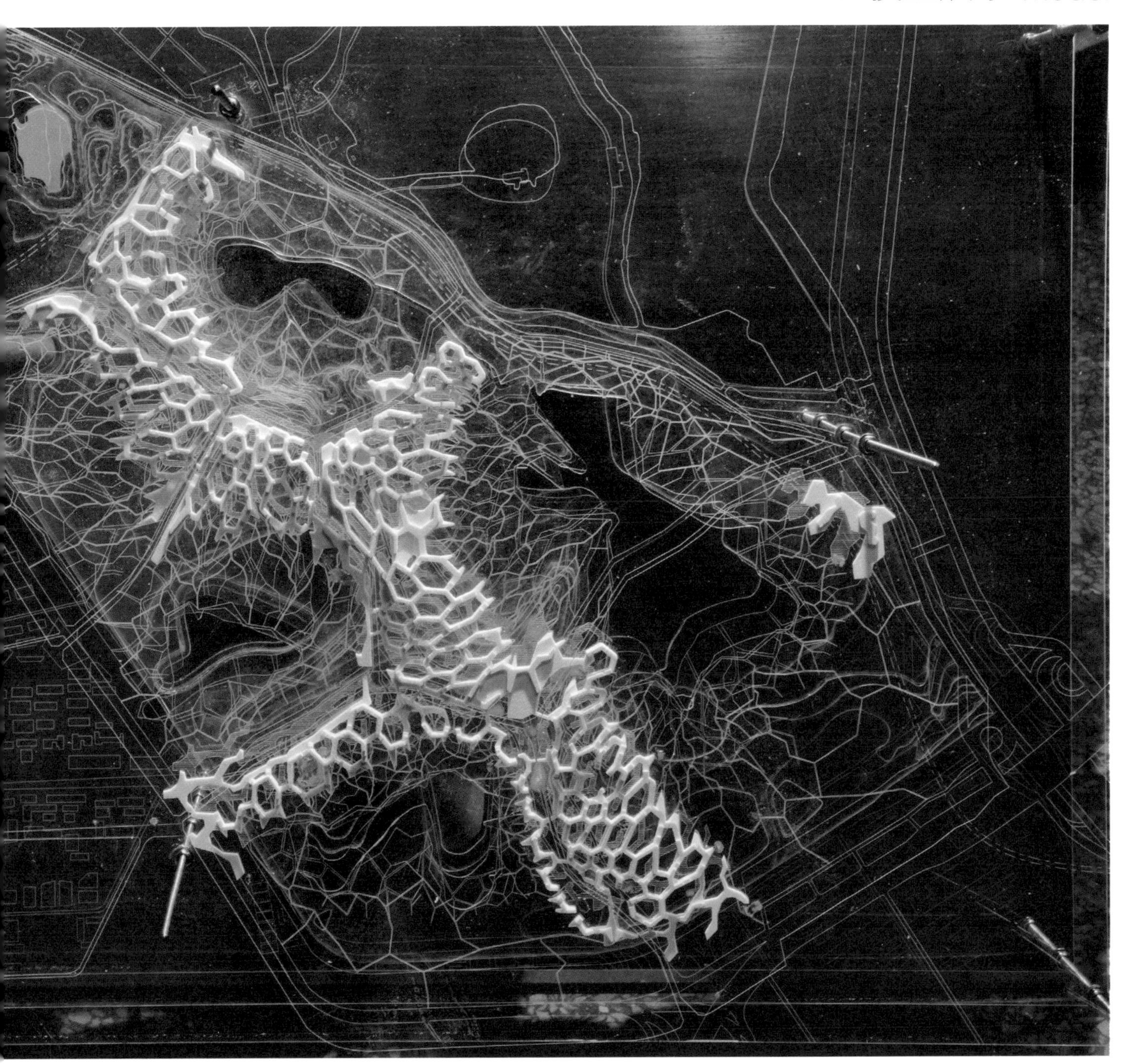

05 模型展示 Model

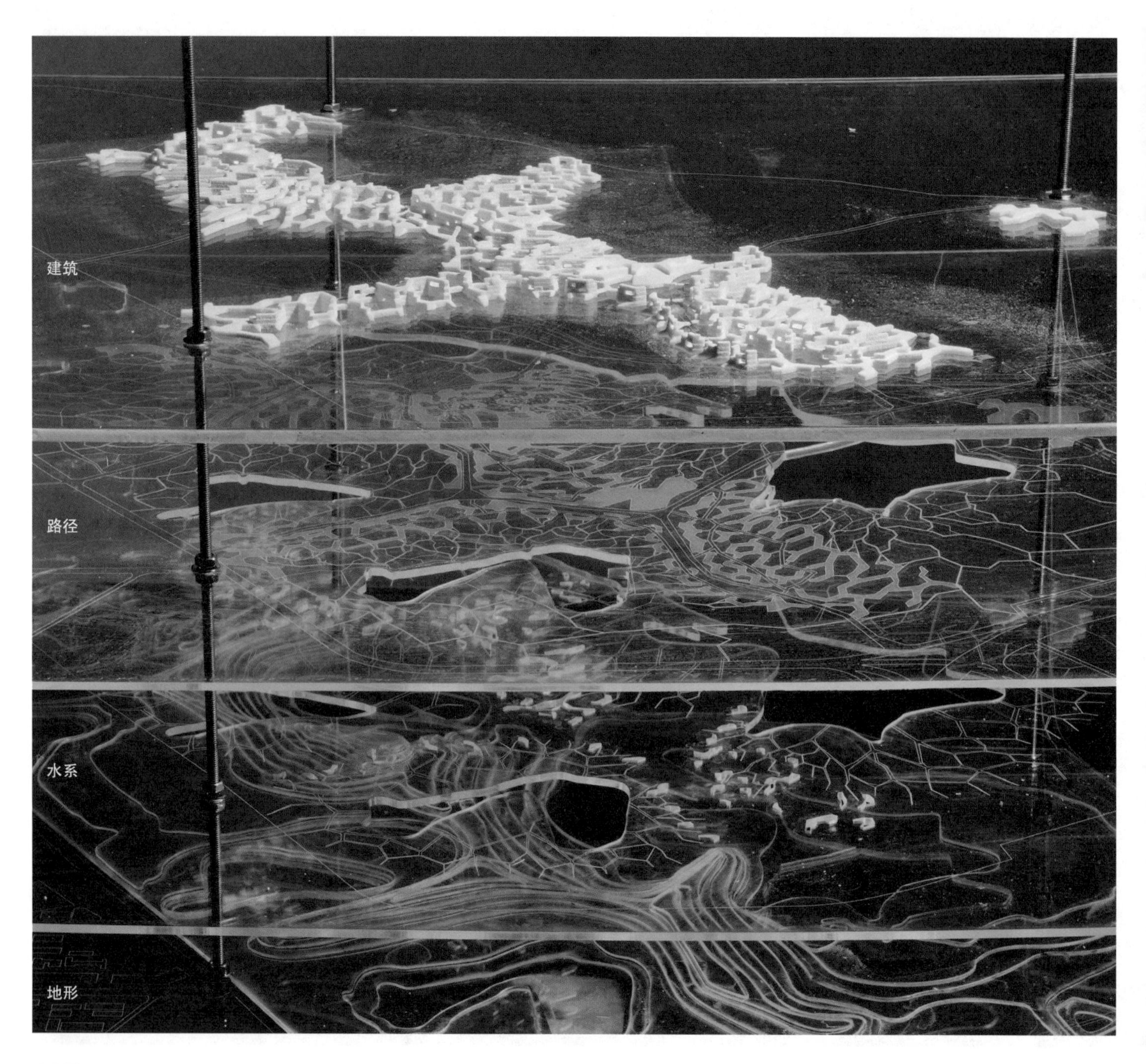

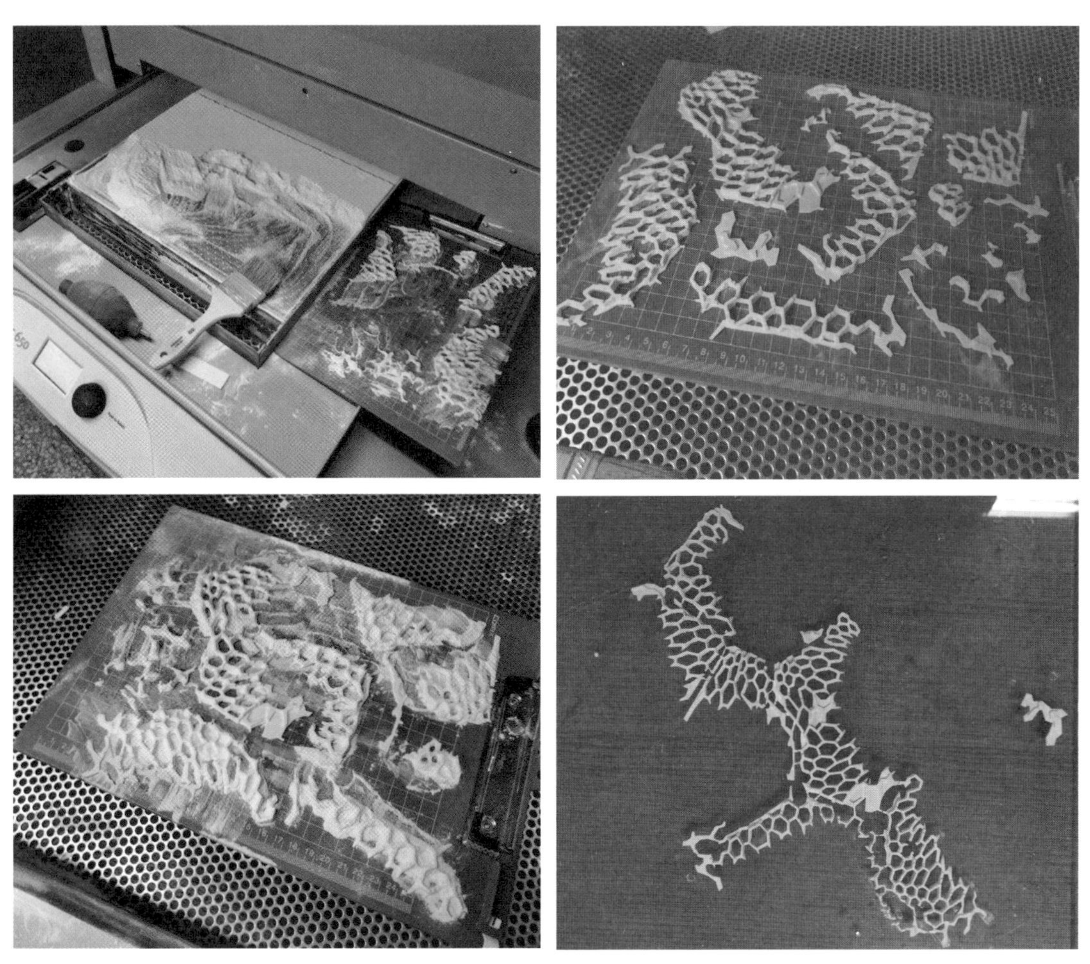

3D 成形研究

日光公园
The Light Park

边思敏
Bian Simin

日光公园是颐和园西南地块的一场光的试验，它抛却了传统的设计方法，力图从更加理性的角度出发，创造这样一座公园：它不仅是一般意义上绿色空间，同时承载了部分城市供电职能；它将市民的休闲活动需求与生产电能的功能需求在同一场地里进行统筹考虑，最终形成了这座以光为主题的公园。

The light park is an experiment located on the southwest of Summer Palace. Different from normal method to design, the light park tried to start in a more rational aspect, to create a space like this: it is not only a traditional green space, but undertake part of electric power supply for the city; it takes both the citizens' needs as a park and the city's need as a part of power system in to consideration . In the end, it comes out as a "light" park.

边思敏

Bian Simin

01 理念构思 Concept

设计概念

场地坐落在北京市的高校和风景名胜集中区——海淀区。场地紧邻颐和园，并且属于北京城市绿带的一部分。为了延续城市绿带，同时保证游客在颐和园中向周围眺望的良好景观视线，场地的使用性质被定义为了一座为市民服务的城市公园。设计关键词有三个：教育性，自给自足，活力。其中“自给自足”是靠昼夜连续工作的太阳能系统实现的。

设计目标是让场地成为一个技术系统和游览系统和谐并存的体系。整个体系由六个高度不等的蓄水湖和贯穿整个场地的台地地形组成。六个蓄水湖两两相邻，一高一低，组成了三个高差蓄水系统。在白天，利用剩余的太阳能将地处的水抽到高处，并在夜晚将水放下以释放电能，供给场地夜晚的用电。台地地形是造景和吸收太阳能的主体部分——平地主要用于公共活动，两层台地间的坡地则主要用于放置太阳能板。两层台地间的坡度通过太阳能吸收率的数据加以分析得到，保证足够的太阳能吸收率。这样一来，既保证了作为城市公园必要的活动和游览空间，又有足够的表面积吸收太阳能，二者互不干扰，并且在一定程度上丰富了公园景观。市民在科技与游览兼具的公园里活动，放松之余也对系统运作有了更深入的了解。

CONCEPT

The site is located in northwest of Beijing, where there are many universities and places of interests, and it is also part of the greenbelt of the city. In order to continue the greenbelt, and not to bring bad view to tourists from Summer Palace as well, the site is designed as a city park where citizens can relax themselves without buying tickets. It has to be educational (like the atmosphere of Haidian District), self-sufficient (to save energy for the whole city), and lively (with various activities in the site).

Since Beijing is a city with abundant sun exposure and insufficient electric energy, this project is supposed to be an ecological park, which absorbs solar power and supports the power consumption of the whole site. Therefore, how to make the technical system and park system work together and create interesting spaces and structures becomes the main target of this project.

To solve the problem above, the structure of the technical system is divided into two parts.The first part is a new topography, pography which is shaped as terraced field. In this way, most of the level ground could be used for activities just like a city park. On the contrary, most of the slopes (and part of level ground) will be filled in solar panels to absorb solar energy for the whole site. The slope of e ach layer is defined by sun exposure absorption.Contour lines with different directions have different slopes to absorb as much solar energy as they can. The second part of the technical system is lakes with difffferent altitudes. This part is used for saving extra solar energy in the daytime by pumping the water from lower lakes to upper lakes, and producting electricity during night by falling down the water from upper lakes to lower lakes. The whole lake system includes six lakes—three higher ones and three lower ones, they save and product energy, and create waterscape for the park at the same time.

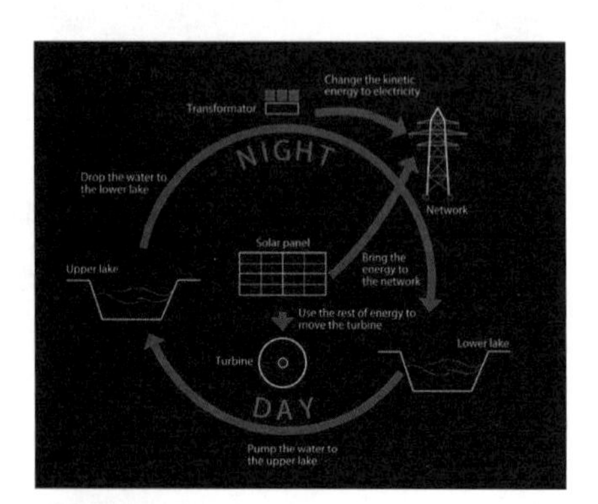

蓄电系统 Energy system

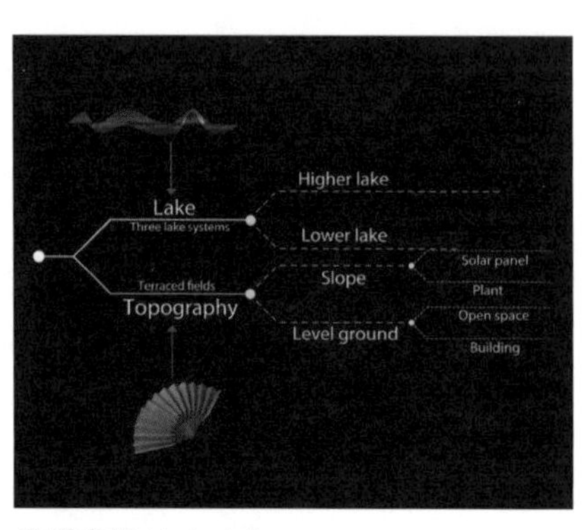

物质空间 Material space

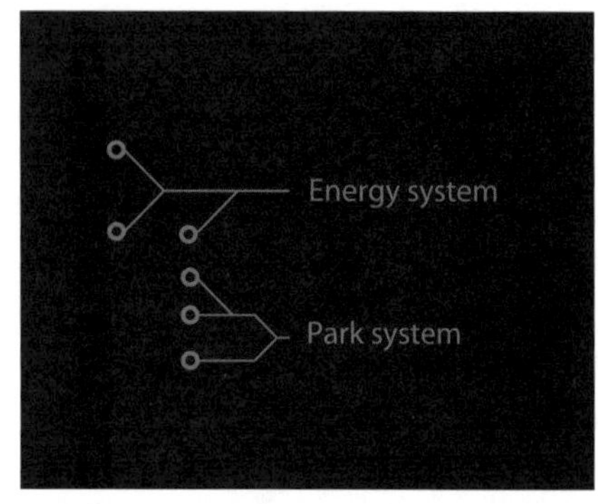

系统分类 System classification

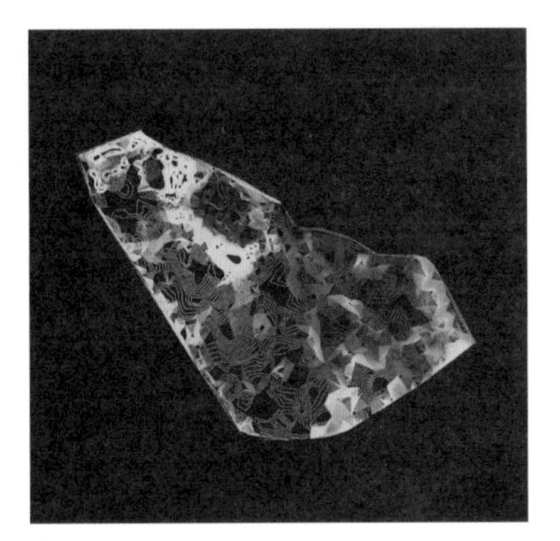

高程 Topography

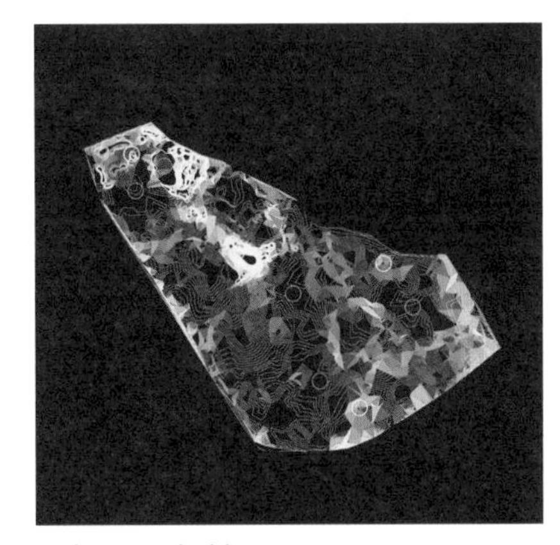

至高 / 至低点选择 Highest & Lowest points

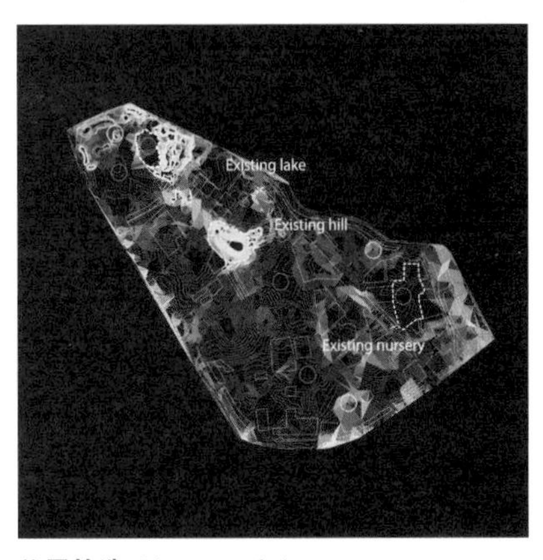

位置筛选 Chose points

蓄水湖最终形态 Final shapes & locations of reservoirs

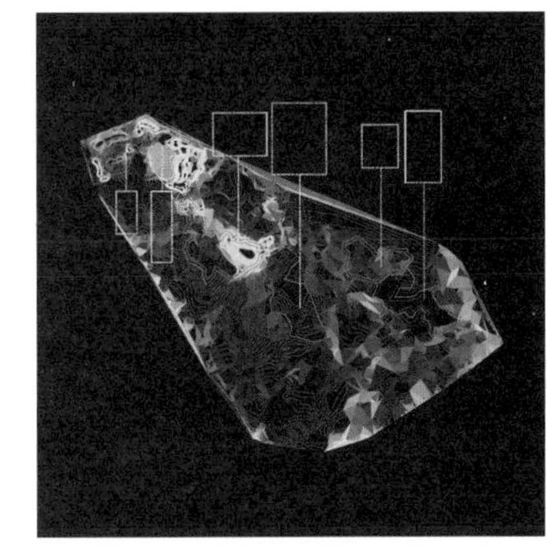

蓄水量分配 Distribution of storage

形态调整 Shape adjustion

02 场地现状 Site Analysis

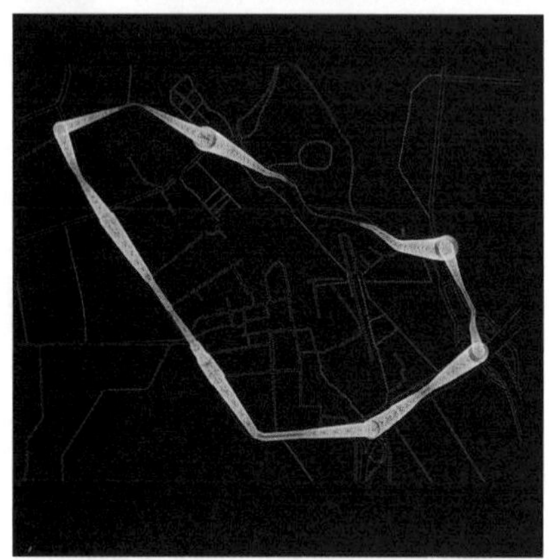
交叉口人流分析 Junction analysis

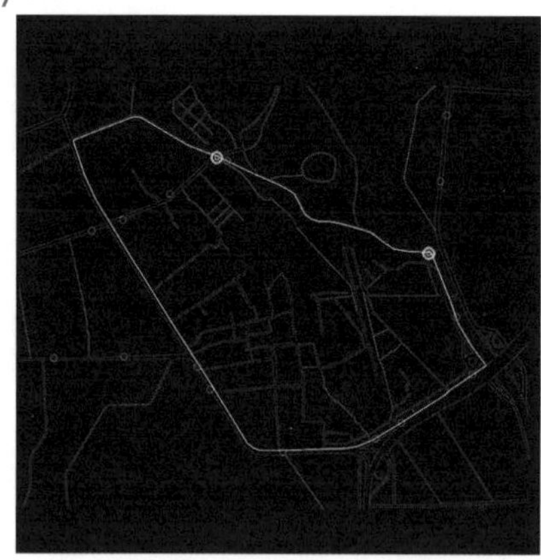
公交及轻轨站点 Bus & tram stations

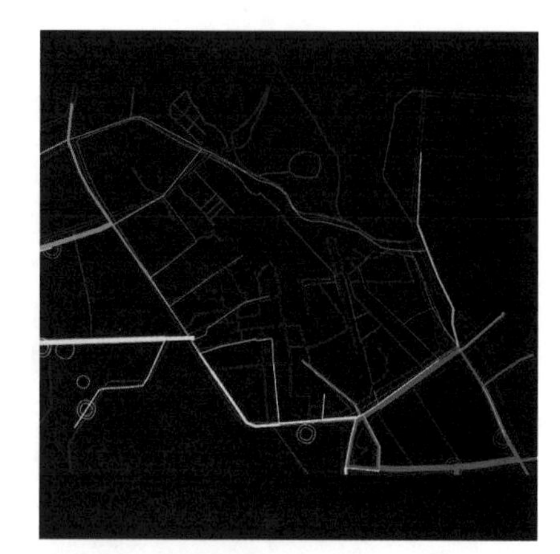
居民区人流分析 Residents analysis

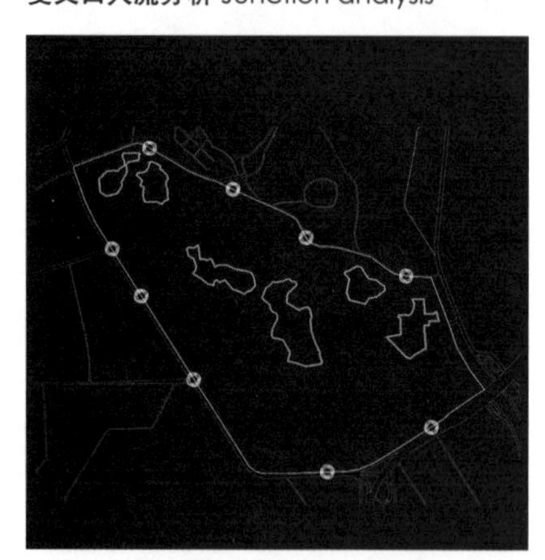
主入口 Main entrances

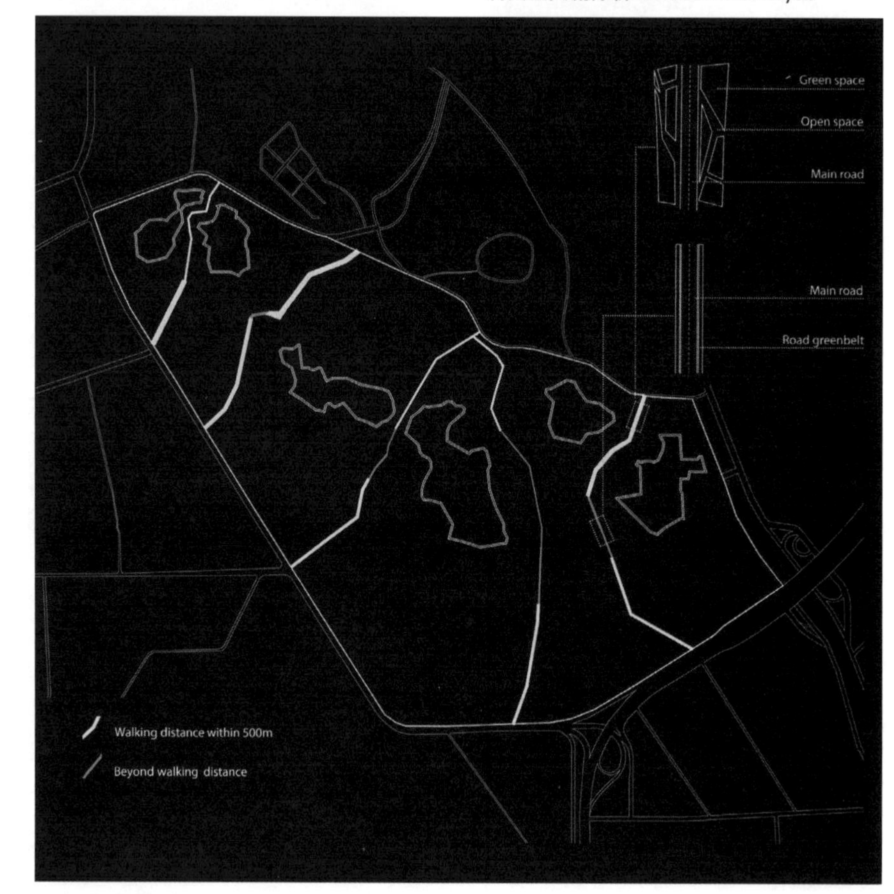

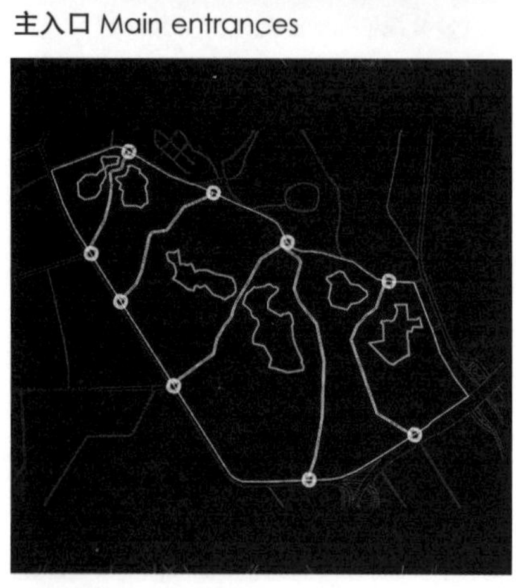
一级道路 Main roads

一级道路宽度确定 Definition of road width

一级道路和蓄水湖 Main roads & reservoirs

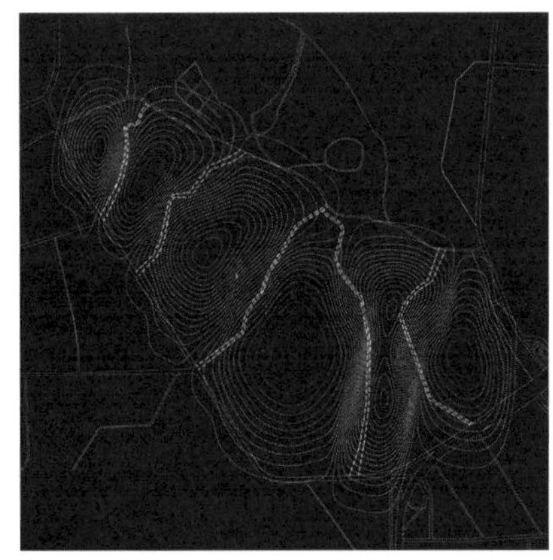

场地高程概念 Concept of topography

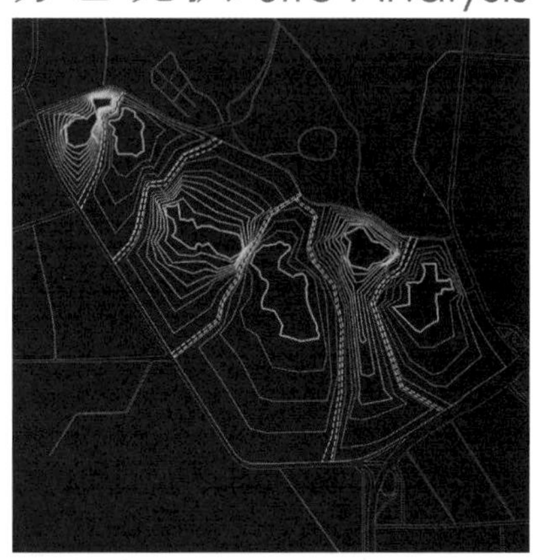

场地新等高线 New topography

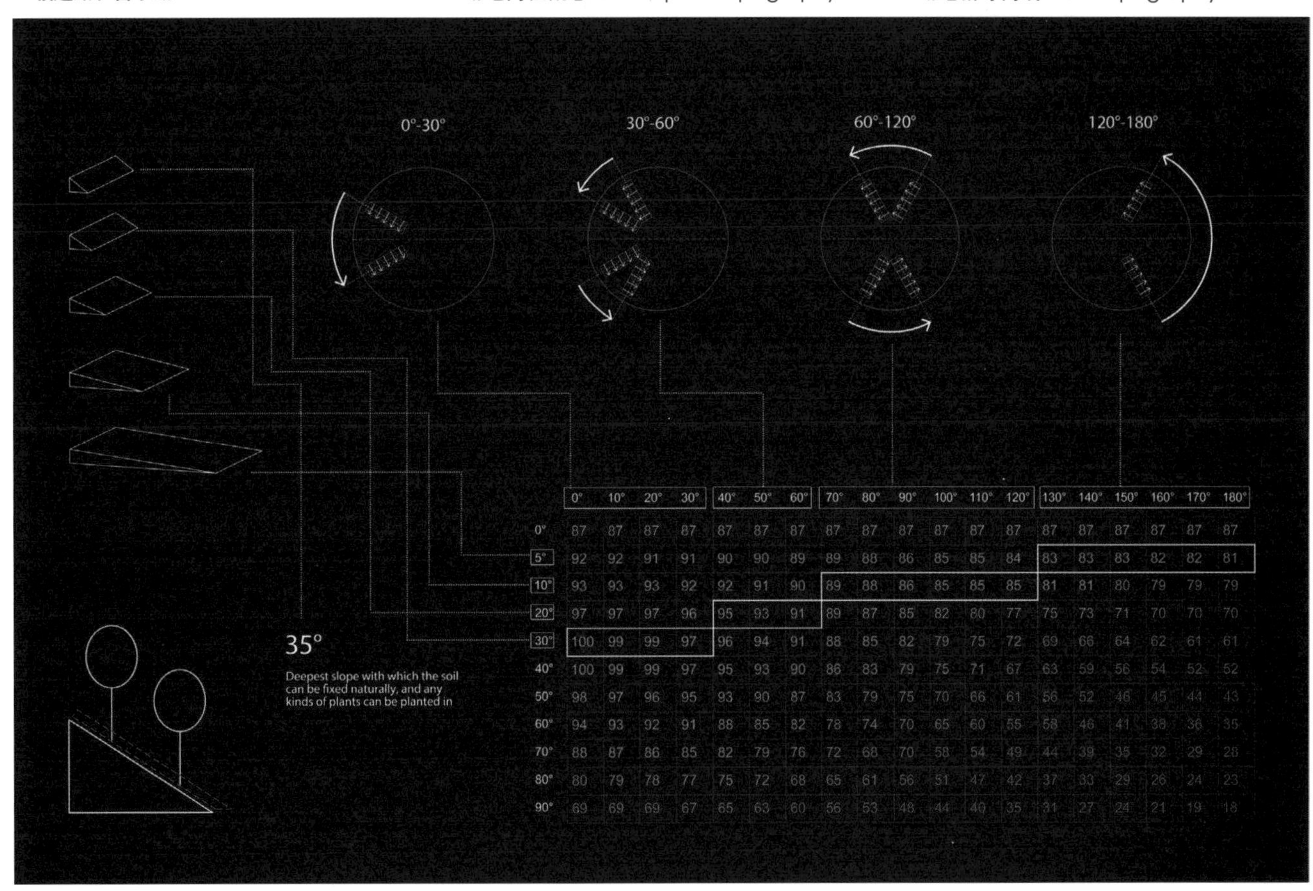

	0°	10°	20°	30°	40°	50°	60°	70°	80°	90°	100°	110°	120°	130°	140°	150°	160°	170°	180°
0°	87	87	87	87	87	87	87	87	87	87	87	87	87	87	87	87	87	87	87
5°	92	92	91	91	90	90	89	89	88	86	85	85	84	83	83	83	82	82	81
10°	93	93	93	92	92	91	90	89	88	86	85	85	85	81	81	80	79	79	79
20°	97	97	97	96	95	93	91	89	87	85	82	80	77	75	73	71	70	70	70
30°	100	99	99	97	96	94	91	88	85	82	79	75	72	69	66	64	62	61	61
40°	100	99	99	97	95	93	90	86	83	79	75	71	67	63	59	56	54	52	52
50°	98	97	96	95	93	90	87	83	79	75	70	66	61	56	52	46	45	44	43
60°	94	93	92	91	88	85	82	78	74	70	65	60	55	58	46	41	38	36	35
70°	88	87	86	85	82	79	76	72	68	70	58	54	49	44	39	35	32	29	28
80°	80	79	78	77	75	72	68	65	61	56	51	47	42	37	33	29	26	24	23
90°	69	69	69	67	65	63	60	56	53	48	44	40	35	31	27	24	21	19	18

等高线分类依据 The principal of the classification of contour lines

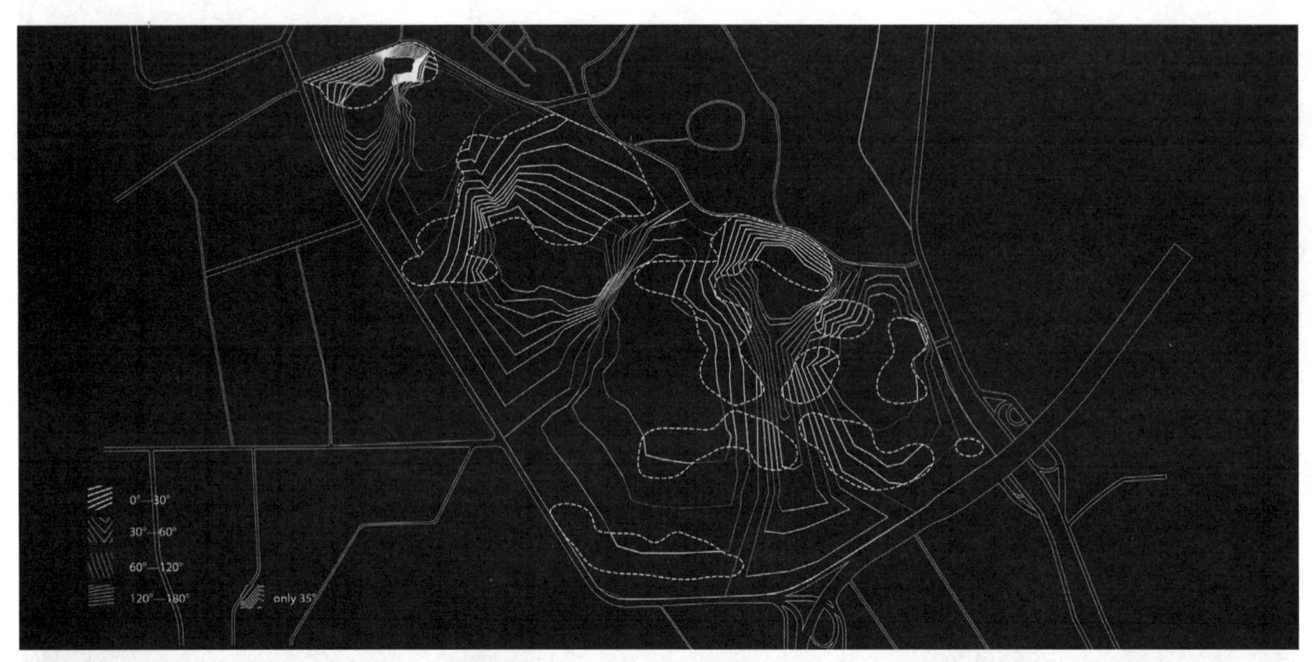

等高线分类 Classification of contour lines

台地地形生成 New terraced field

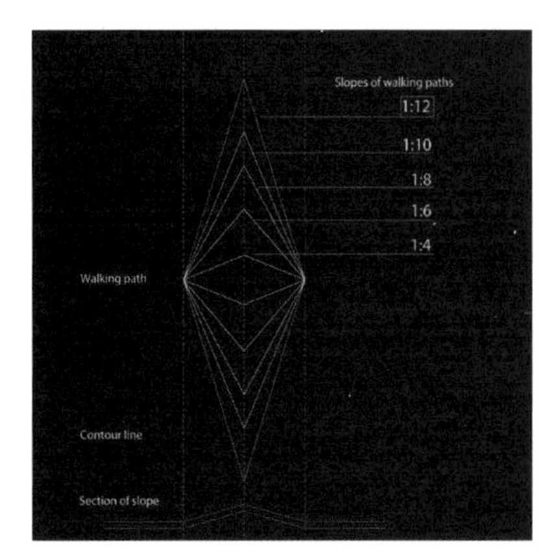

坡度分析 Slope analysis

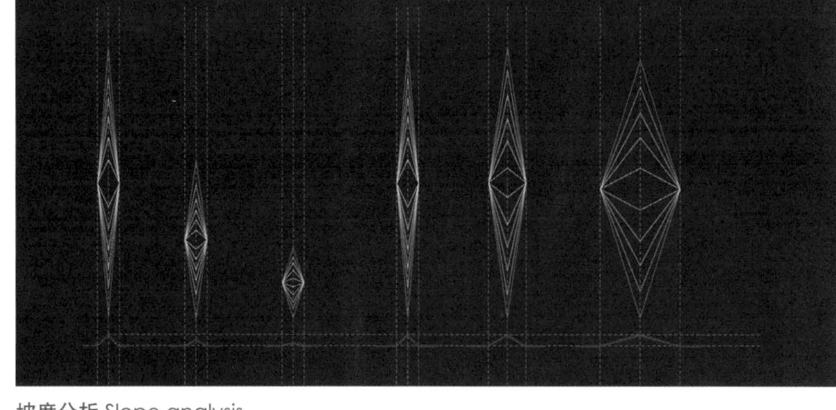
坡度分析 Slope analysis

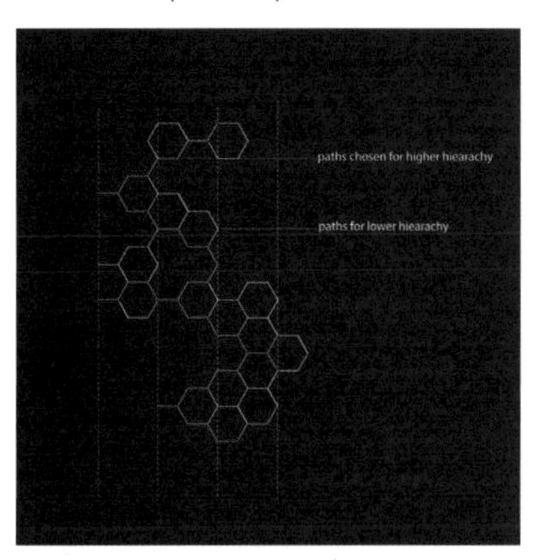

等级分类 Hierarchy Classification

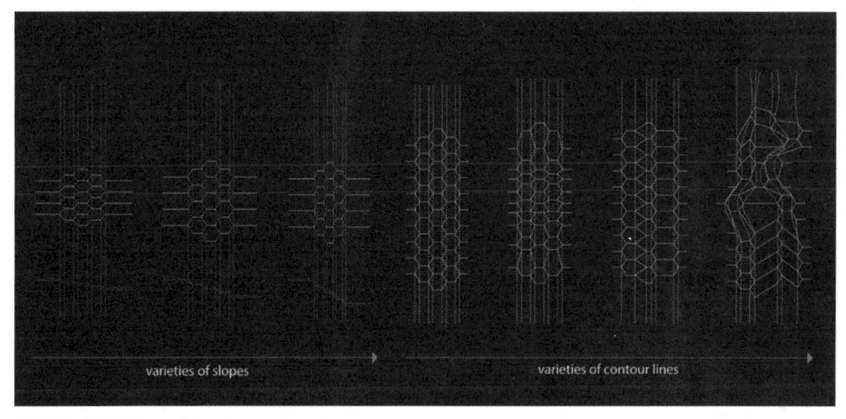

等高线变化 Diversity of contour lines

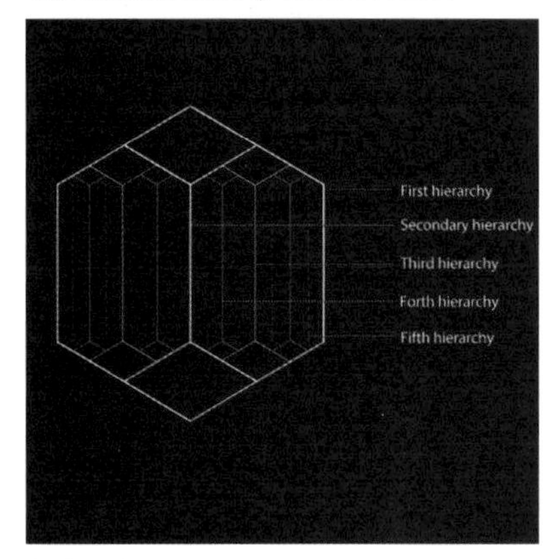

细胞细分 Sub division

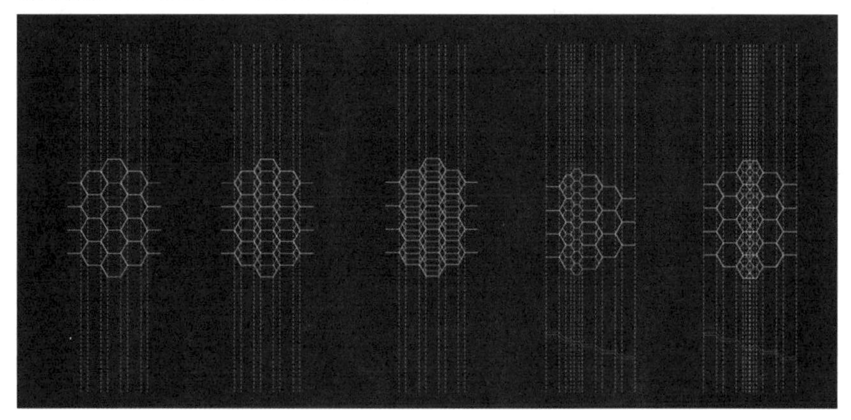
细胞变化 Diversity of cells

04 详细设计 Detailed Design

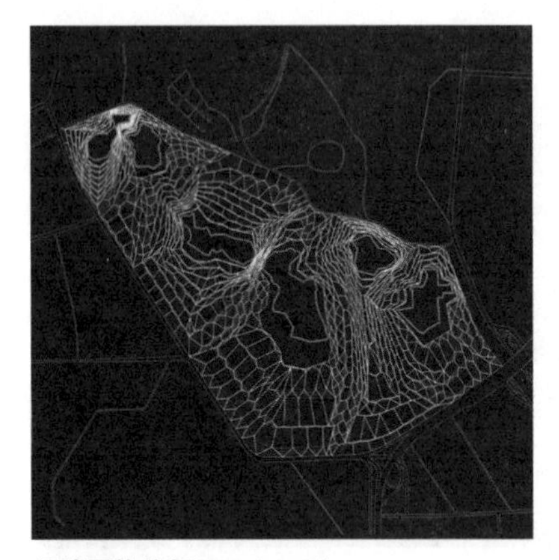

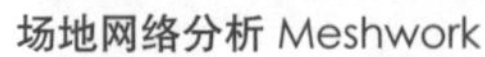

场地网络分析 Meshwork

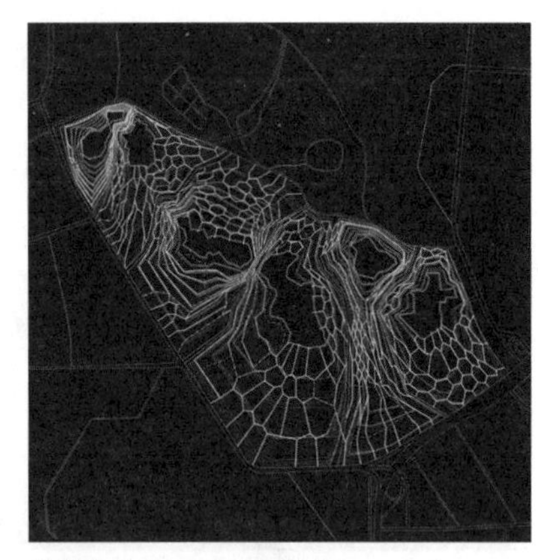

道路选线 Chose roads from mesh

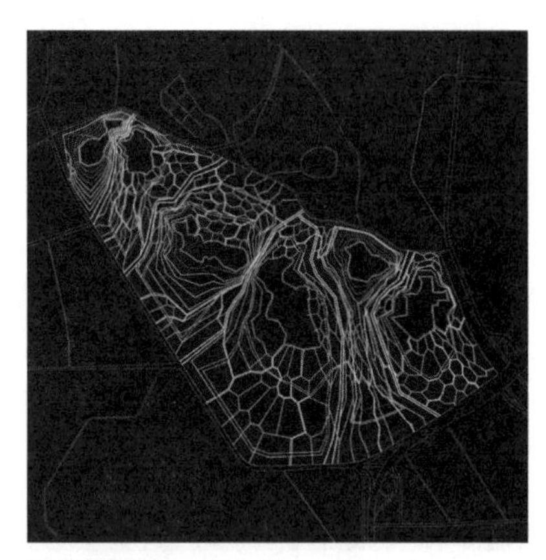

道路等级 Classification of roads

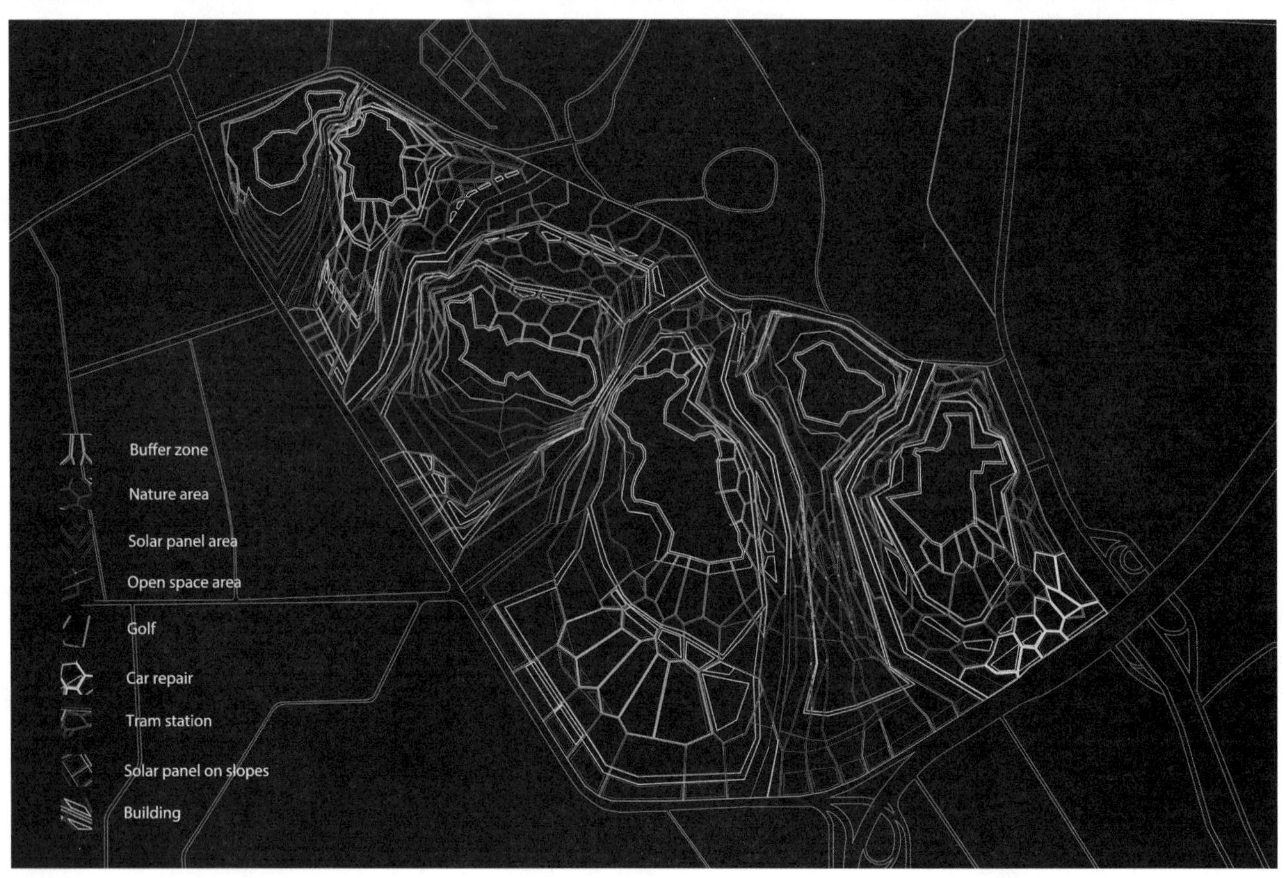

功能分布 Distribution of functions

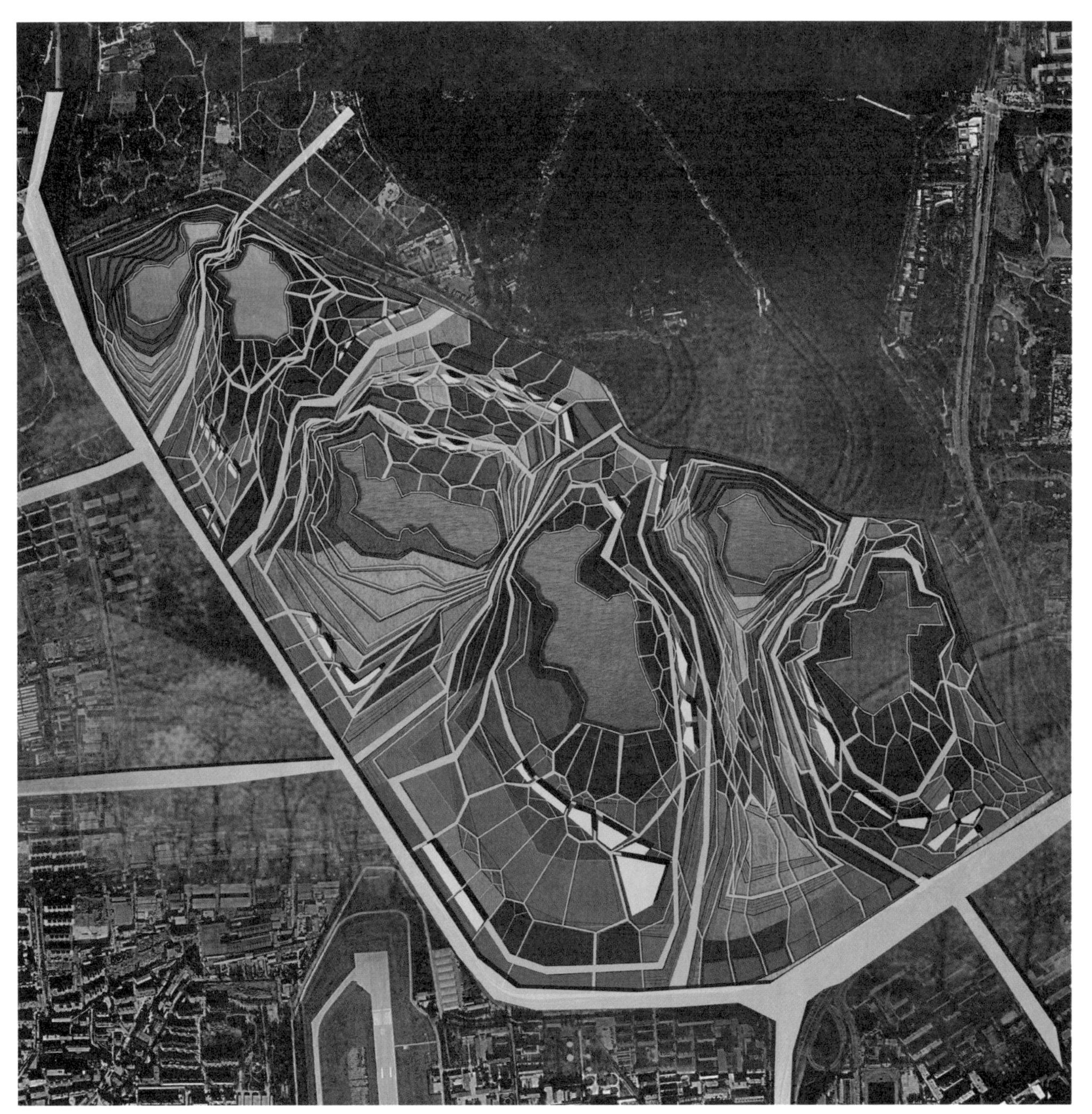

总平面图 Master plan

透视图 1 Rendering1

透视图 2 Rendering2

带状绿色基础设施

Strip Under Strip

朱一君
Zhu Yijun

本案通过对地形的改造来重建一套生态基础设施。这套基础设施不仅可以作为雨洪状态下的洪水通道也可以作为正常状态下的雨水收集和净化系统。
为了实现以上目标，本案提出了一种高效的树形水系统作为整个地景改造的基底。树形水系统的主干作为中水和雨水收集的通道，而“枝杈”部分则主要用于洪水的疏通。通过对几何特征和材料组织的研究，这套树形系统发展出一套具体而落地的形式单元并通过同构的重复和变形形成一个网状控制系统。

This project, on territorial scale, proposes to construct a topography-based “landscape infrastructure” to respond to multiple challenges. The overarching concept of the design is to build topography as emergency flood container and daily rainwater collecting corridor, and as grey water treatment and collecting corridor as well, dividing a variety of landscape functions, compressing substrate soil for development.
To achieve this goal, an efficient branching water system is designed in the first place. The system following water flow logic consists of two parts, one is for daily rainwater and grey water, and another sub-branching is for emergency flood.

朱一君

Zhu Yijun

01 场地现状 Site Analysis

BACKGROUND AND CONTEXT

With the development of human construction and excess water consumption, Haidian District in Beijing, which used to be known as its farmland and wetland, is losing the unique water landscape.

The site locates between 4th Ring and 5th Ring Road, and the borders northeast with Summer Palace of high historic value, south with a military airport and residential, and west with some office buildings and residential.

This is a territory of rich history but of relatively new formation. However, more and more residents are moved away from this area, because of the construction of a reservoir, which serves the south-north project. This area, once a beautiful farmland and wetland landscape, is facing a trouble to become a wasteland.

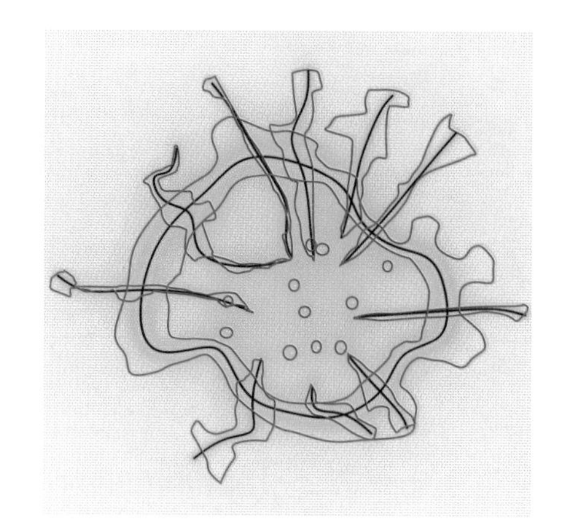

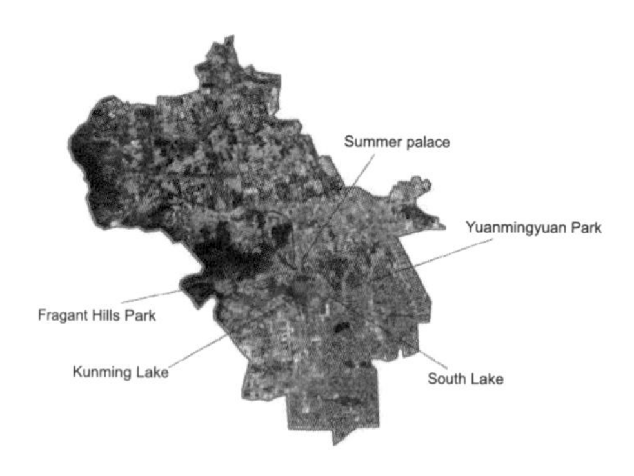

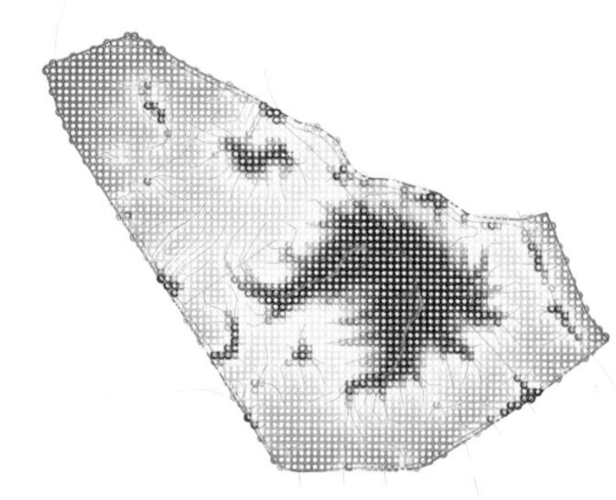

CHANLLEGE

Beijing is one of the driest capitals in the world; however, it has been suffered from sudden flood in recent years, which mainly results from its inefficient water system. Beijing is facing a water crisis!

The underground water level becomes lower, even the mother river of ancient Beijing has almost disappeared. Furthermore, lots of housing, especially in suburban area, are exposed to flood risks, rendering the site almost impossible for any future development.

WORKING PRECESS

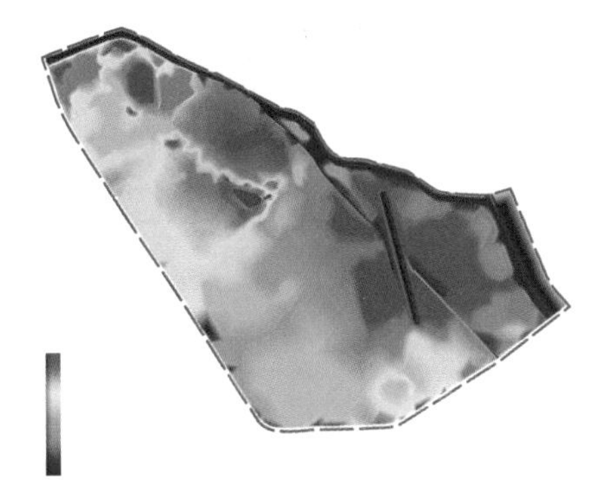

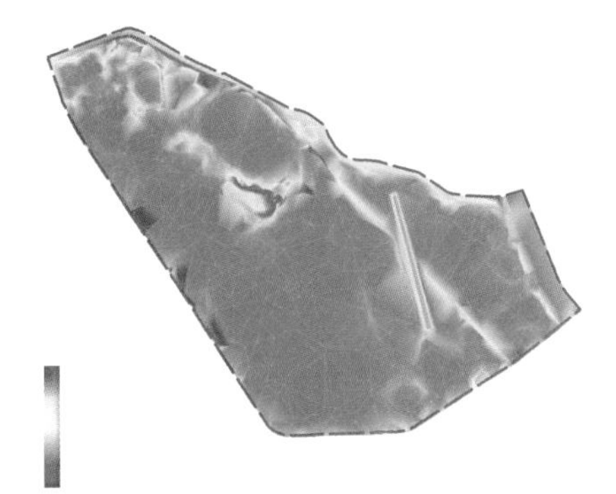

02 理念构思 Concept

DOMESTIC WATER CONSUMPTION	$44m^3$/person/year
grey water consumption (75%)	$33m^3$/person/year
black water consumption (25%)	$11m^3$/person/year
WATER CONSUMPTION IN OFFICES	$6m^3$/person/year
AGRICULTURE WATER USE	1500
PARK WATER USE	1500
WATER RECYCLING	$3m^3$/person

Residents	1000
Housing floor are	40000☒
Wetland	3000☒
Re-using water in building (25% of water use)	$11000m^3$、year
Green land	7hectares

OPPORTUNITY AND OBJECTIVE

Could there be a way to convert site's decline, even recover its water landscape? Furthermore, could there be a way to drain the flood and reuse it for daily life of the surrounding housing? How to make full use the grey water of the territory and its periphery? How to create a park with proper urbanization and let the city and green space serve each other? The project strives to answer the foregoing questions and brings them into synergy, by recognizing the structural force of landscape in the transformation process that manages to construct an environmental network on a territory scale.

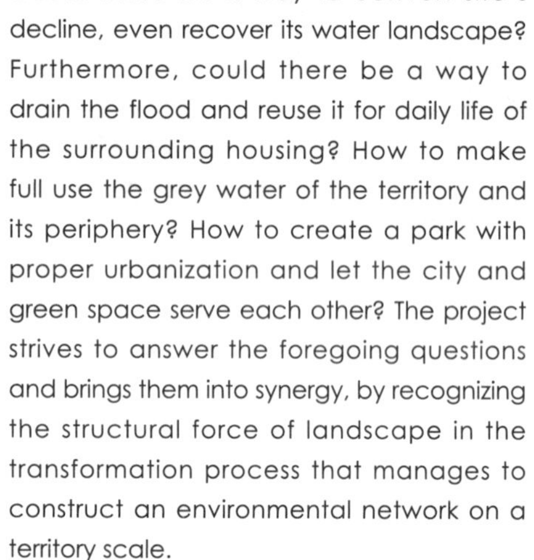

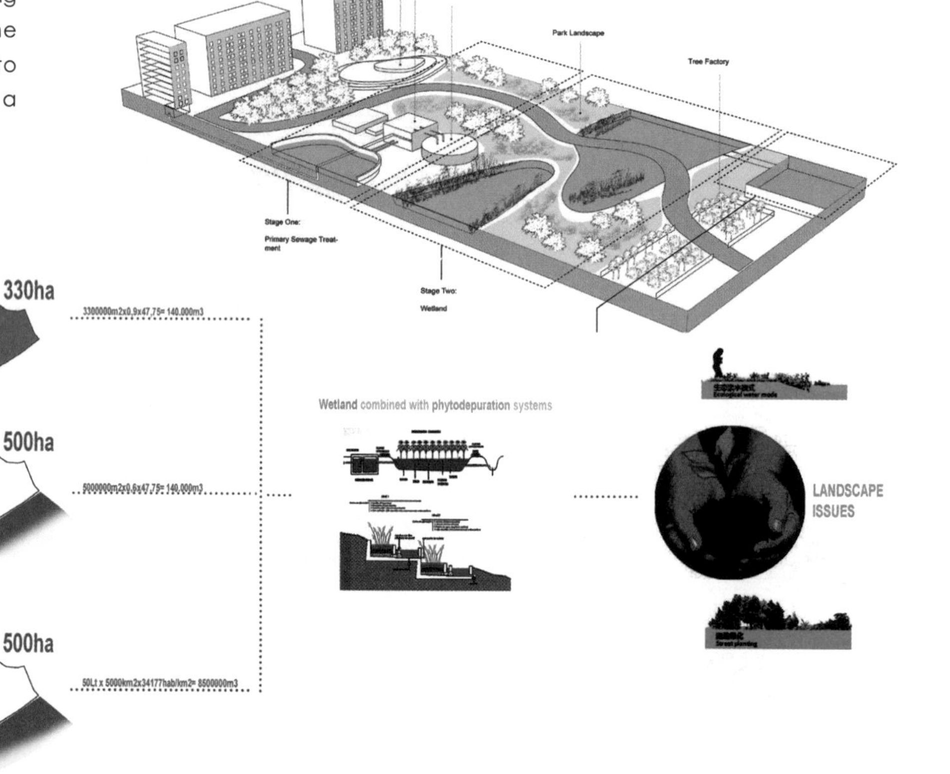

SOLUTION AND DESIGN

This project, on territorial scale, proposes to construct a topography-based "landscape infrastructure" to respond to multiple challenges. The overarching concept of the design is to build topography as emergency flood container and daily rainwater collecting corridor, and as grey water treatment and collecting corridor as well, dividing a variety of landscape functions, compressing substrate soil for development. To achieve this goal, an efficient branching water system is designed in the first place. The system following water flow logic consists of two parts, one is for daily rainwater and grey water, and another sub-branching is for emergency flood.

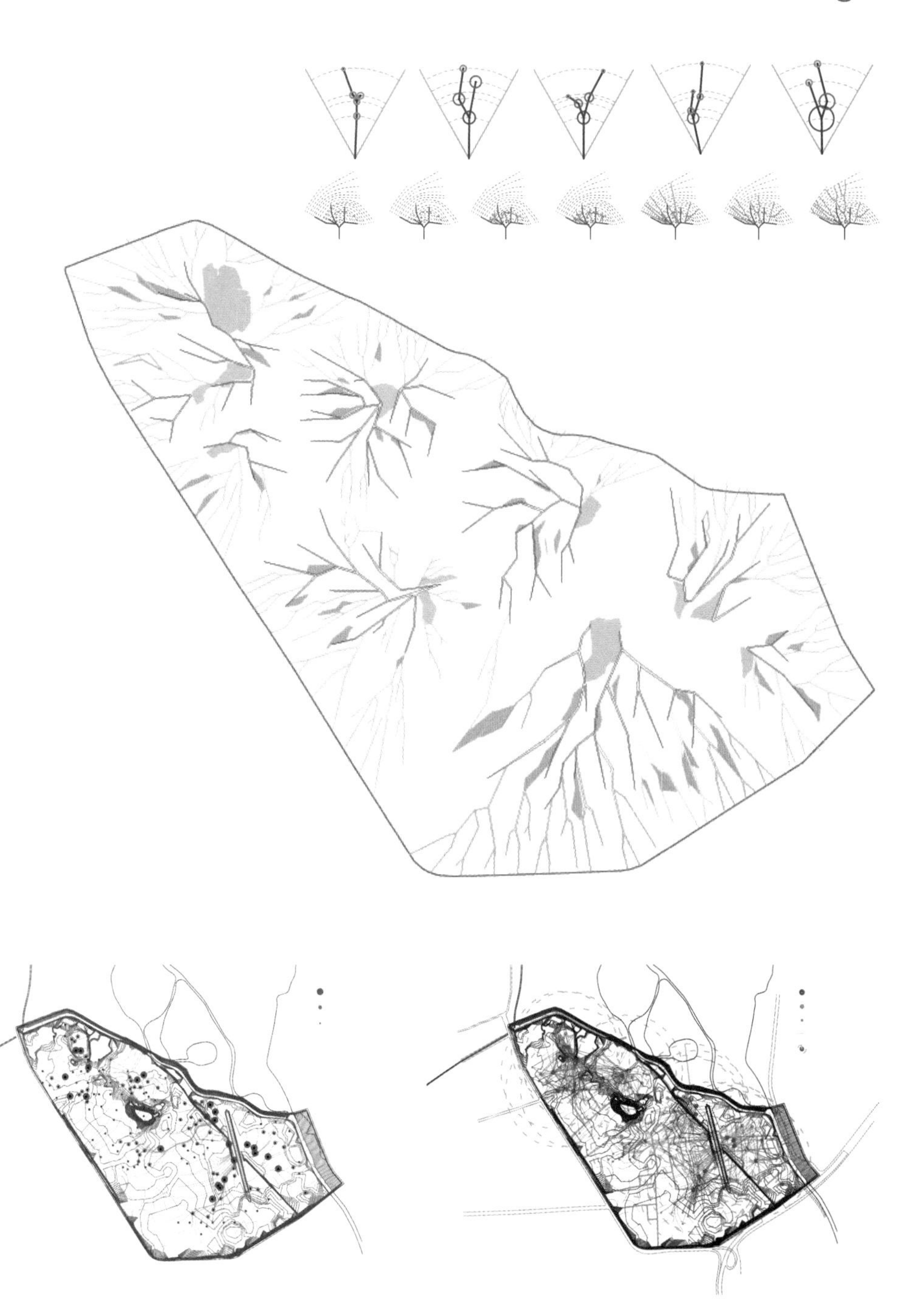

03 总体设计 General Design

SOLUTION AND DESIGN

A new topography is generated by the following rules: 1.identifying water body outside the site as future outlet; 2.defining wetland of collecting as topographical low points; 3.inserting collecting water canals into a variety of landscape function; 4.creating corridors connecting existing urban fabric and new development by sloping towards low points.

This landscape infrastructure accommodates flexible utilization patterns, in terms of both different scenarios and varied water volumes, generating rich gradients of interaction between fixed ground and dynamic natural forces.

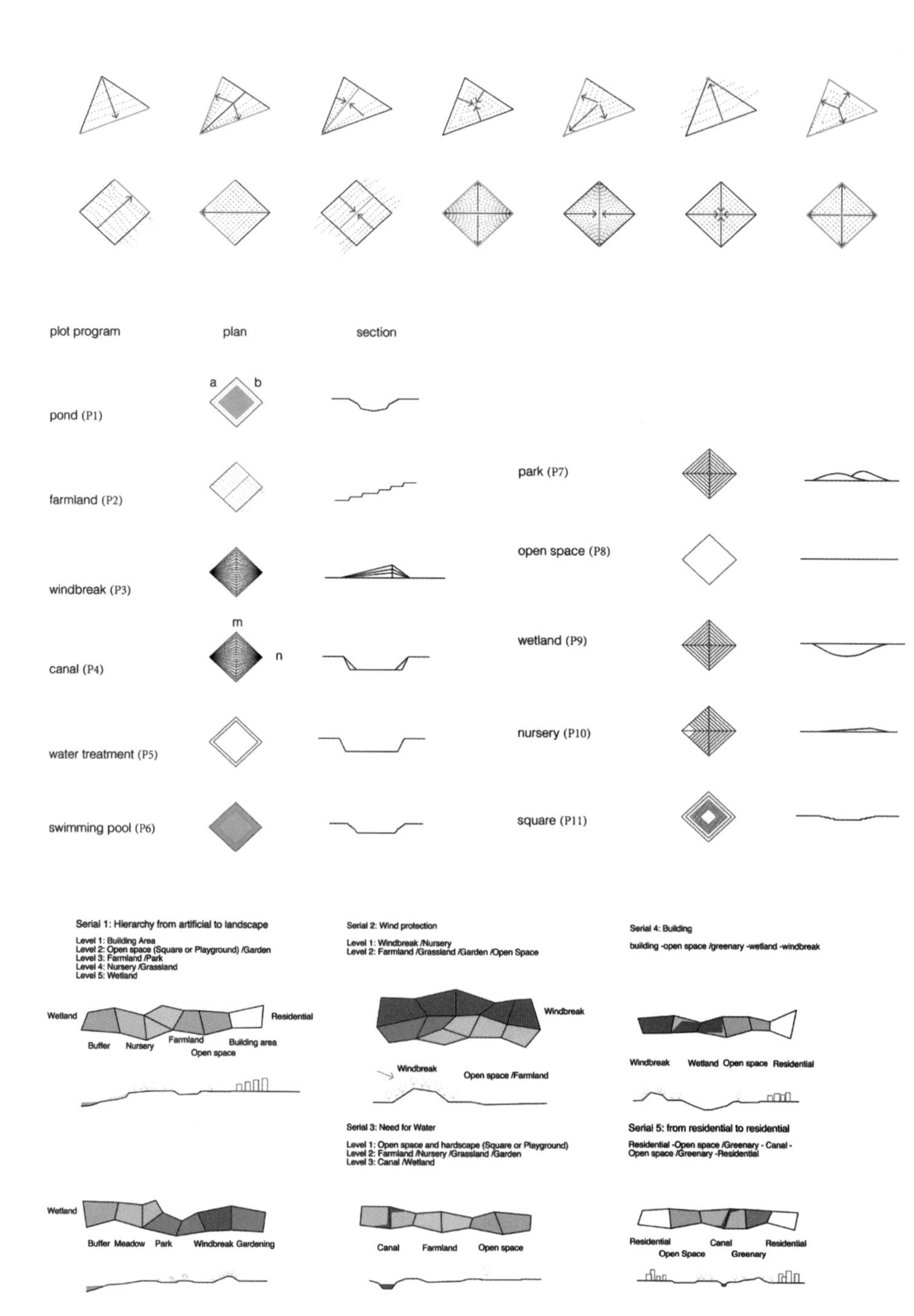

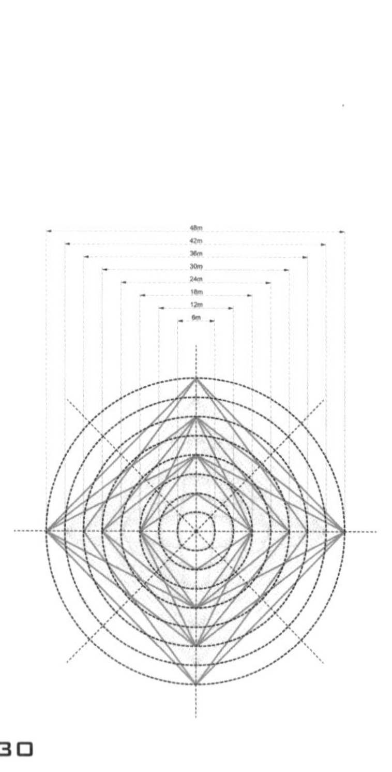

After a research on the geometry feature and material organization of component, the branching system generates a new mesh network based on components.

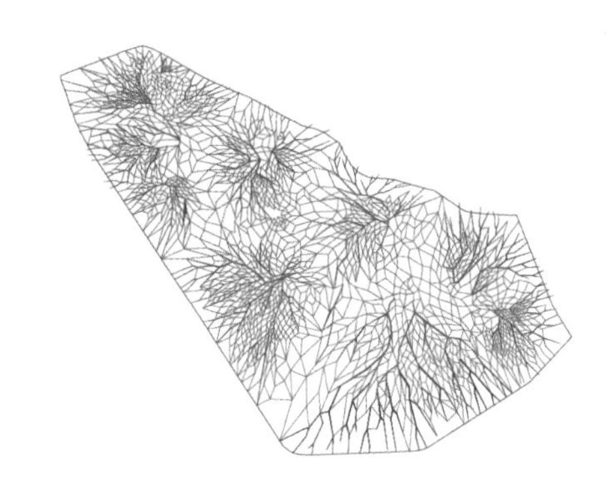

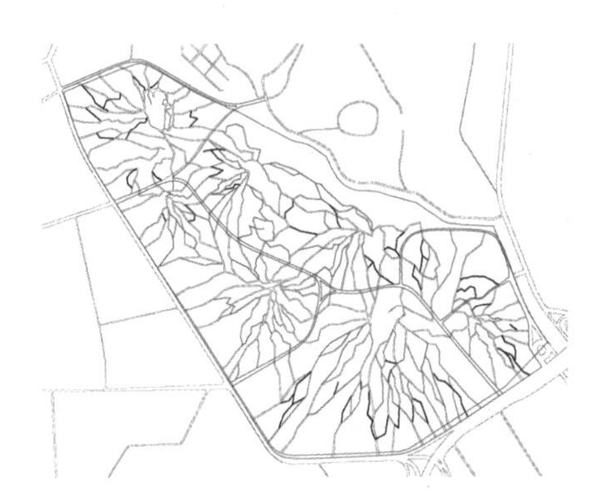

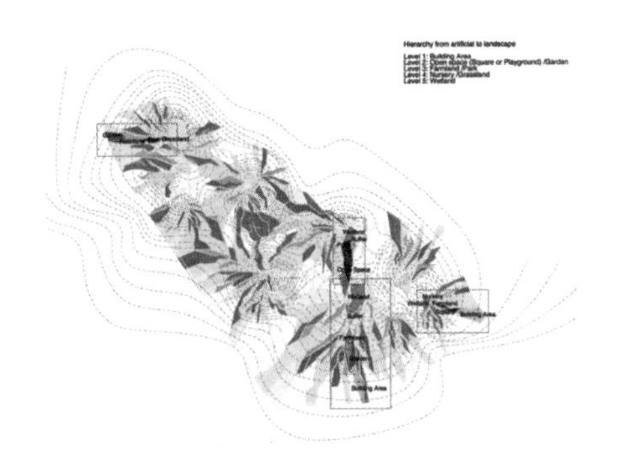

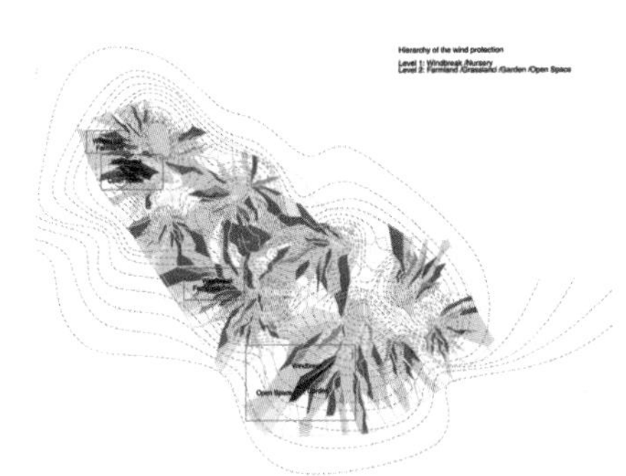

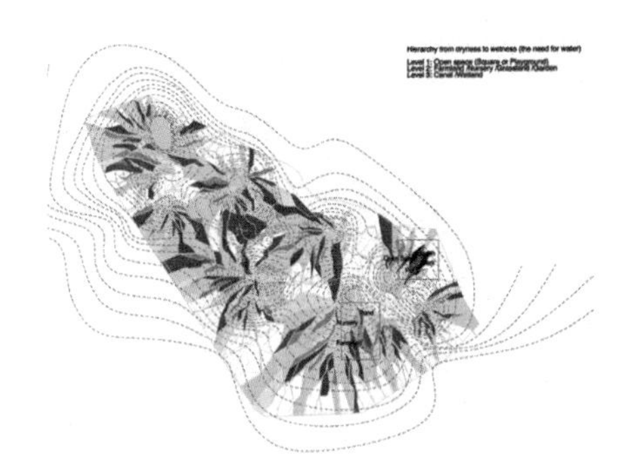

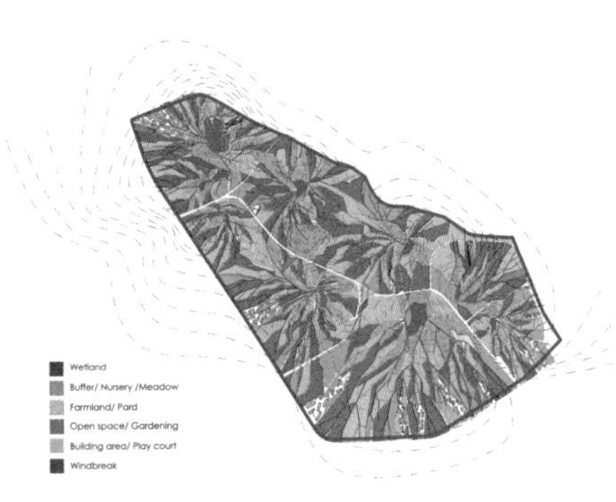

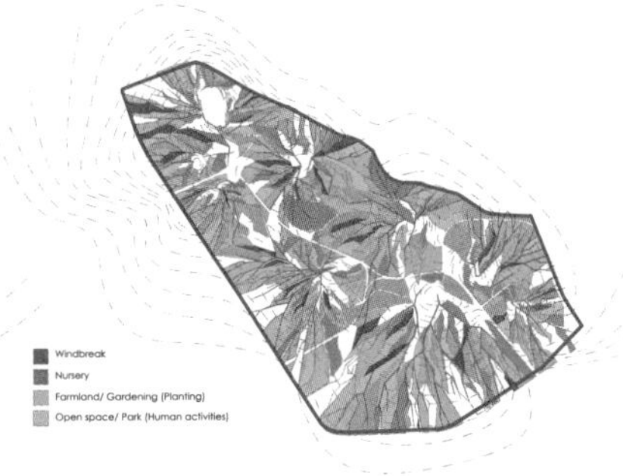

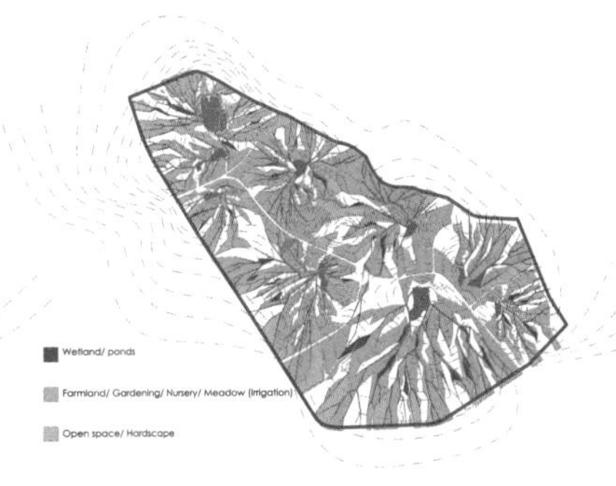

Thus, rich gradients of landscape patterns are presented in the territory. The main logic for the program hierarchy comes from a desire to have a smooth transition from artificial to natural. The protection by wind barriers, the adjacency to water body and the relationship between different urban areas are also applied in the program arrangement. A lot of sections are designed to illustrate the relationship of these program serials.

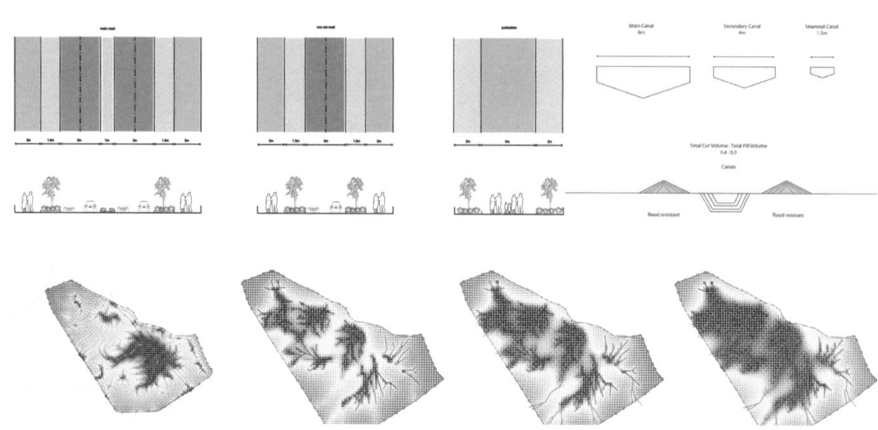

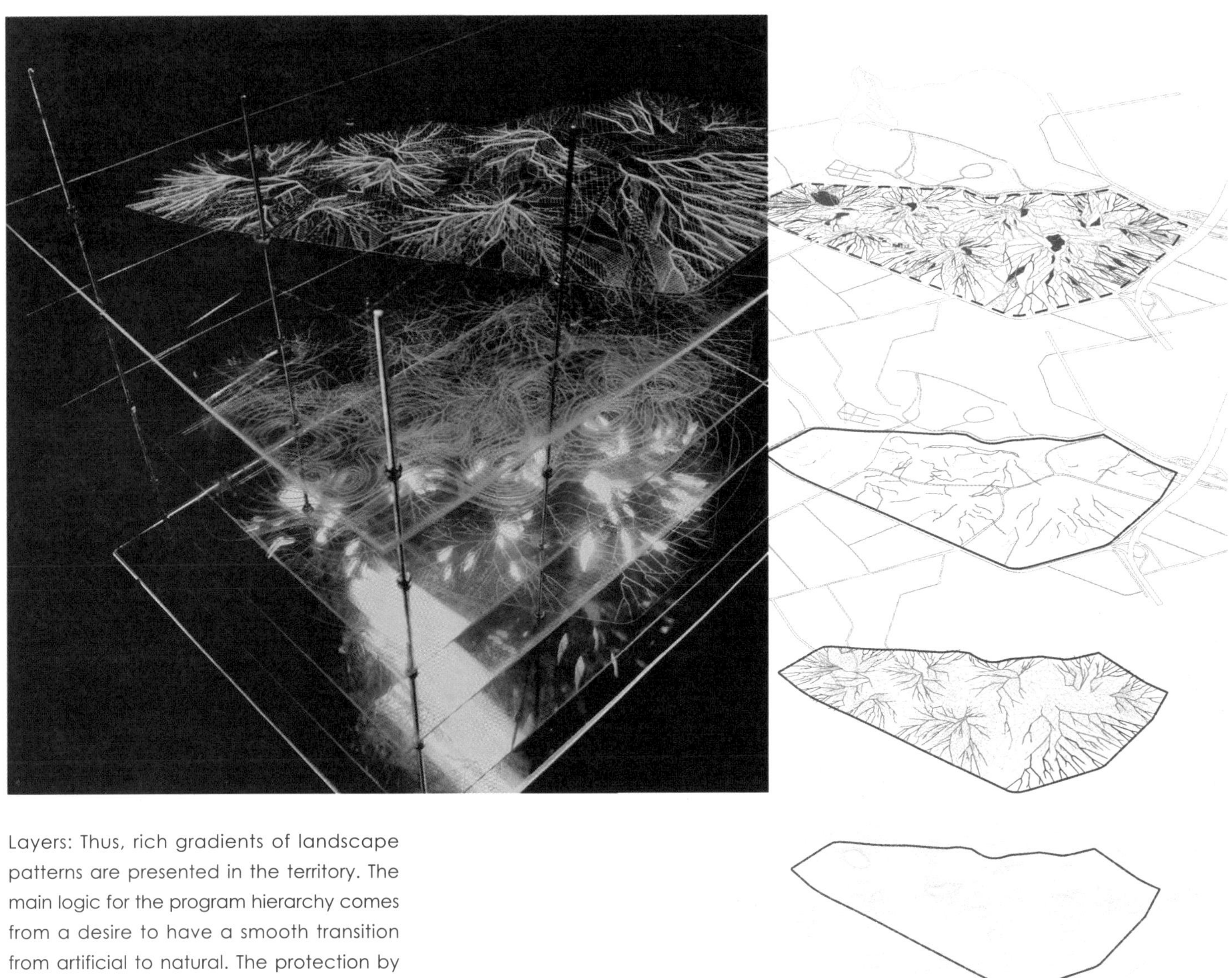

Layers: Thus, rich gradients of landscape patterns are presented in the territory. The main logic for the program hierarchy comes from a desire to have a smooth transition from artificial to natural. The protection by wind barriers, the adjacency to water body and the relationship between different urban areas are also applied in the program arrangement. A lot of sections are designed to illustrate the relationship of these program serials.

04 详细设计 Detailed Design

Thus, rich gradients of landscape patterns are presented in the territory. The main logic for the program hierarchy comes from a desire to have a smooth transition from artificial to natural. The protection by wind barriers, the adjacency to water body and the relationship between different urban areas are also applied in the program arrangement. A lot of sections are designed to illustrate the relationship of these program serials.

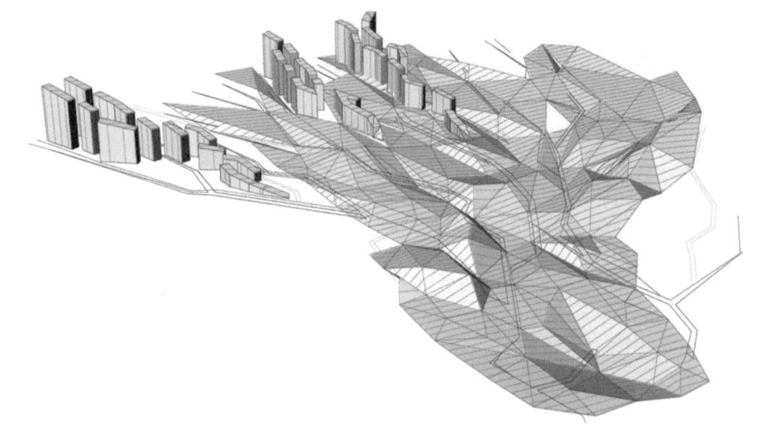

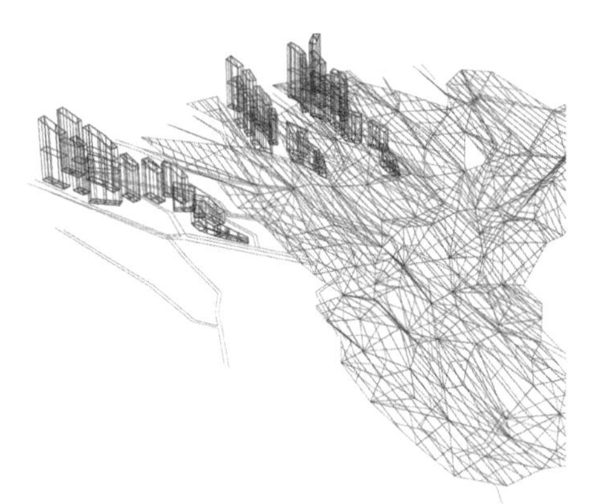

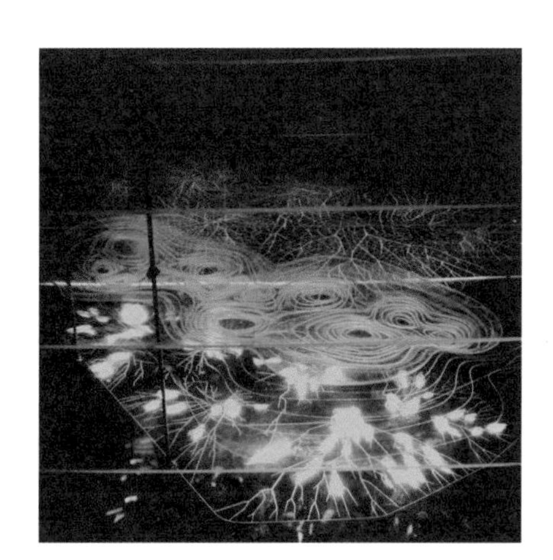

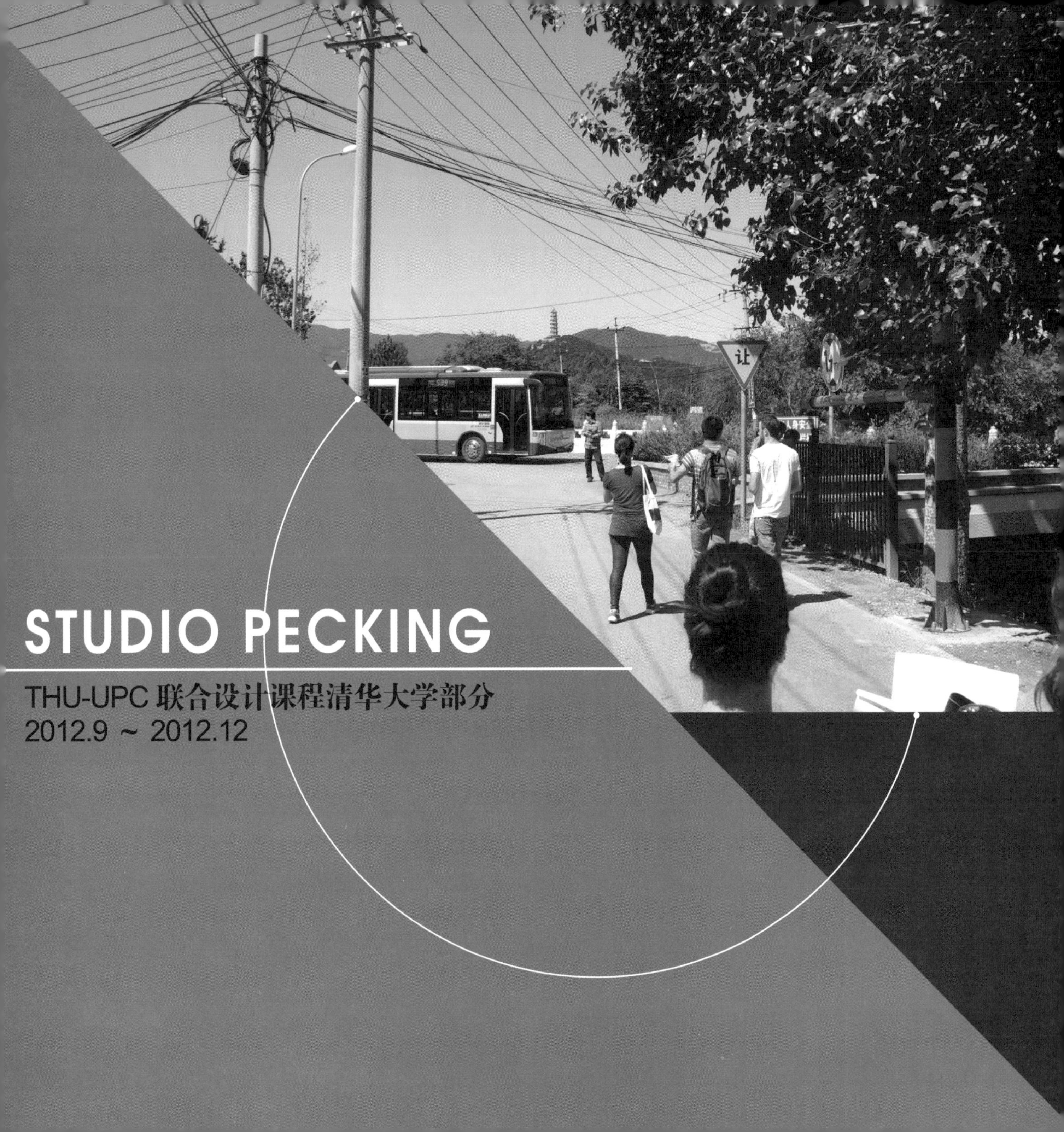

STUDIO PECKING

THU-UPC 联合设计课程清华大学部分

2012.9 ~ 2012.12

记·载

For memory

梁思佳　冯阳　罗茜
Liang Sijia/Feng Yang/Luo Qian

设计的出发点在于对现状两条轴线的理解与处理。作为南水北调工程的一个重要端头，现状的水渠在整个场地中占有极其重要的位置。形式上，这条轴线限定了整个场地的空间格局和属性，也是设计无论如何都要予以尊重和保留的。原有的金河导流渠体现了现状场地的机理和空间秩序，但它同样也是阻碍整个场地更新发展的一个重要的元素。因此，新的设计保留了现有的南水北调水渠，并改造了现有的金河导流渠，并试图通过改变这两条轴线的相互关系而重塑场地的空间结构。

The design concept based on the understanding and attitude of the two axes on the site. As the end of the south-to-north water diversion project, the canal takes an extremely important position in the entire site.

Formally, this axis defines the spatial pattern and properties of the entire site. Therefore the canal in this case must be respected and retained. Although the original Jin River diversion canal reflects the spatial order of the site, but it is also an important element to obstruct the updating and developing of the entire site.

01 理念构思 Concept

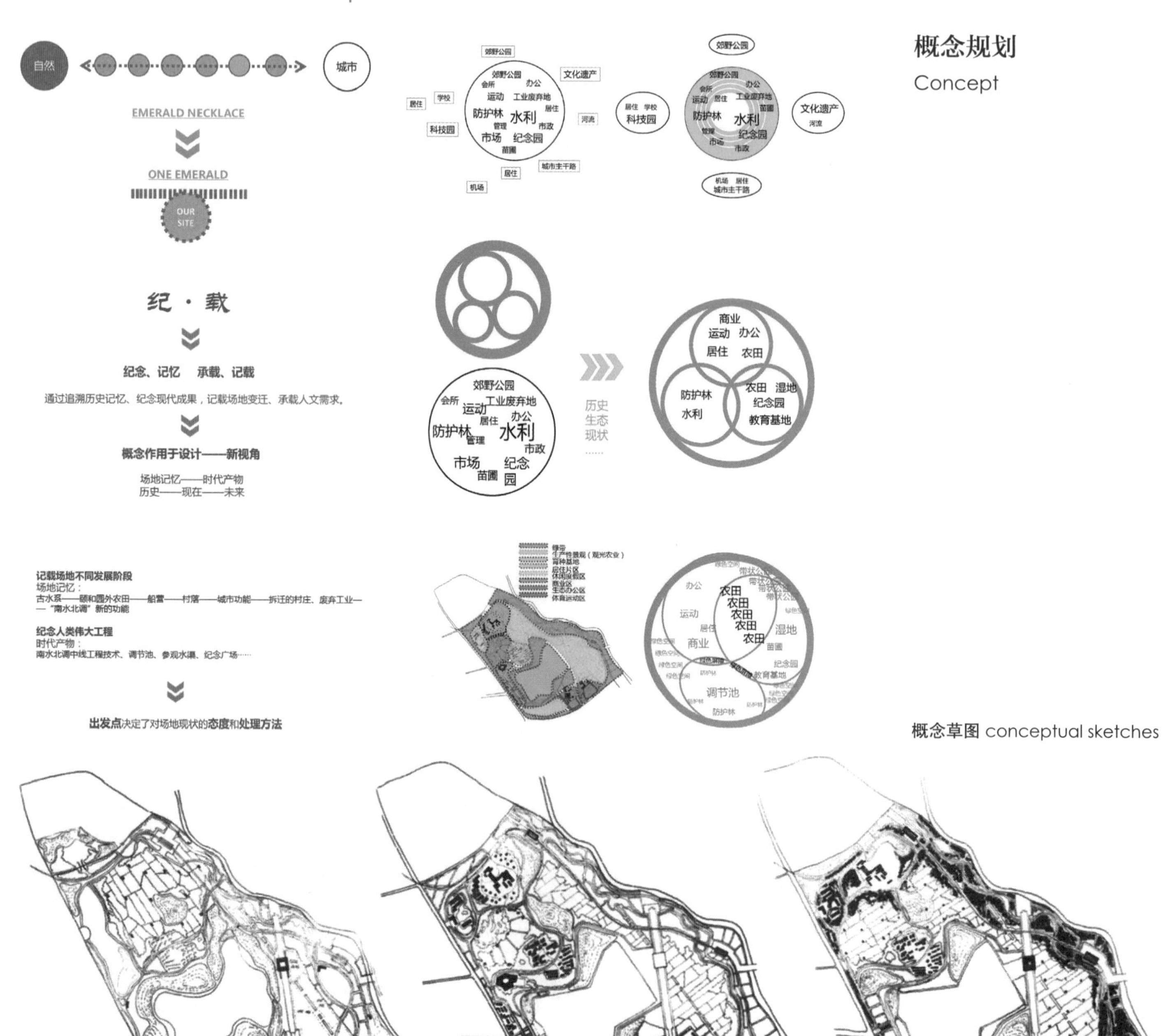

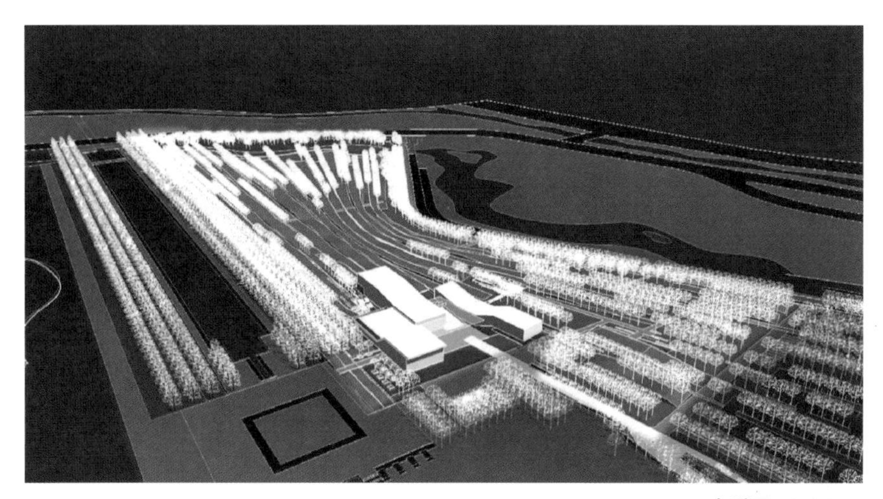

鸟瞰图 birdview

设计的出发点在于对现状两条轴线的理解与处理。作为南水北调工程的一个重要端头，现状的水渠在整个场地中占有极其重要的位置。形式上，这条轴线限定了整个场地的空间格局和属性，也是设计无论如何都要予以尊重和保留的。原有的金河导流渠体现了现状场地的机理和空间秩序，但它同样也是阻碍整个场地更新发展的一个重要的元素。因此，新的设计保留了现有的南水北调水渠，并改造了现有的金河导流渠，并试图通过改变这两条轴线的相互关系而重塑场地的空间结构。

因此，金河导流渠的改线就成为进一步设计的必然，在新旧两条轴线之间的场地则成为设计的重点。我们试图找寻原有轴线之间的形式联系与韵律。意图创造富有情趣和意味的园林纪念空间。

大地艺术的理念和形式给了我们设计的灵感。线性的平面布局形式成为我们解决平面结构问题的方式。我们希望通过起伏的地形，有明确指向性的树阵作为场地的结构，通过强烈的形式语言将我们的概念发展到场地中去。我们弱化了原有宏大轴线的端头，将出入口的位置进行了调整。我们试图通过一套可行的形式语言发展出满足场地功能需求的设计。

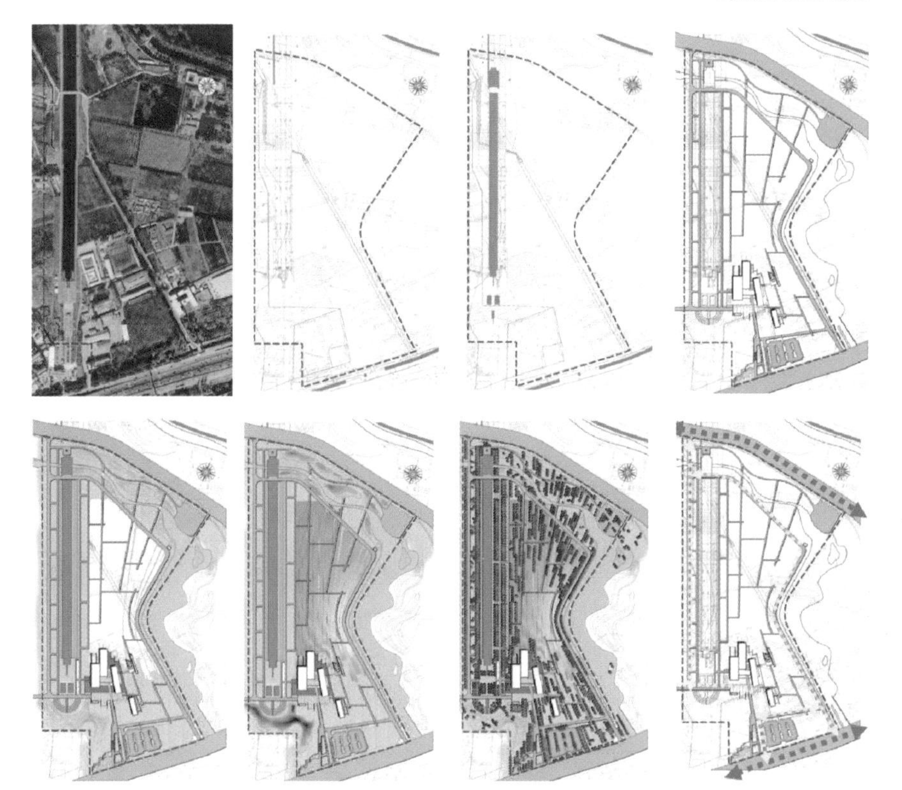

概念推演及分析 concept deduce and analysis

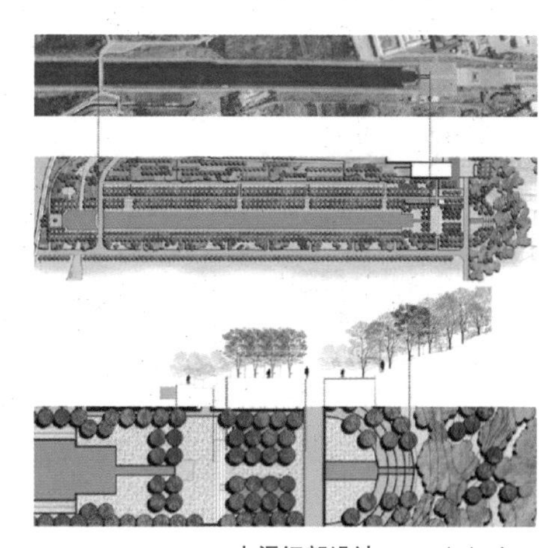

水渠细部设计 canal design

02 总体设计 General Design

设计思路
Idea

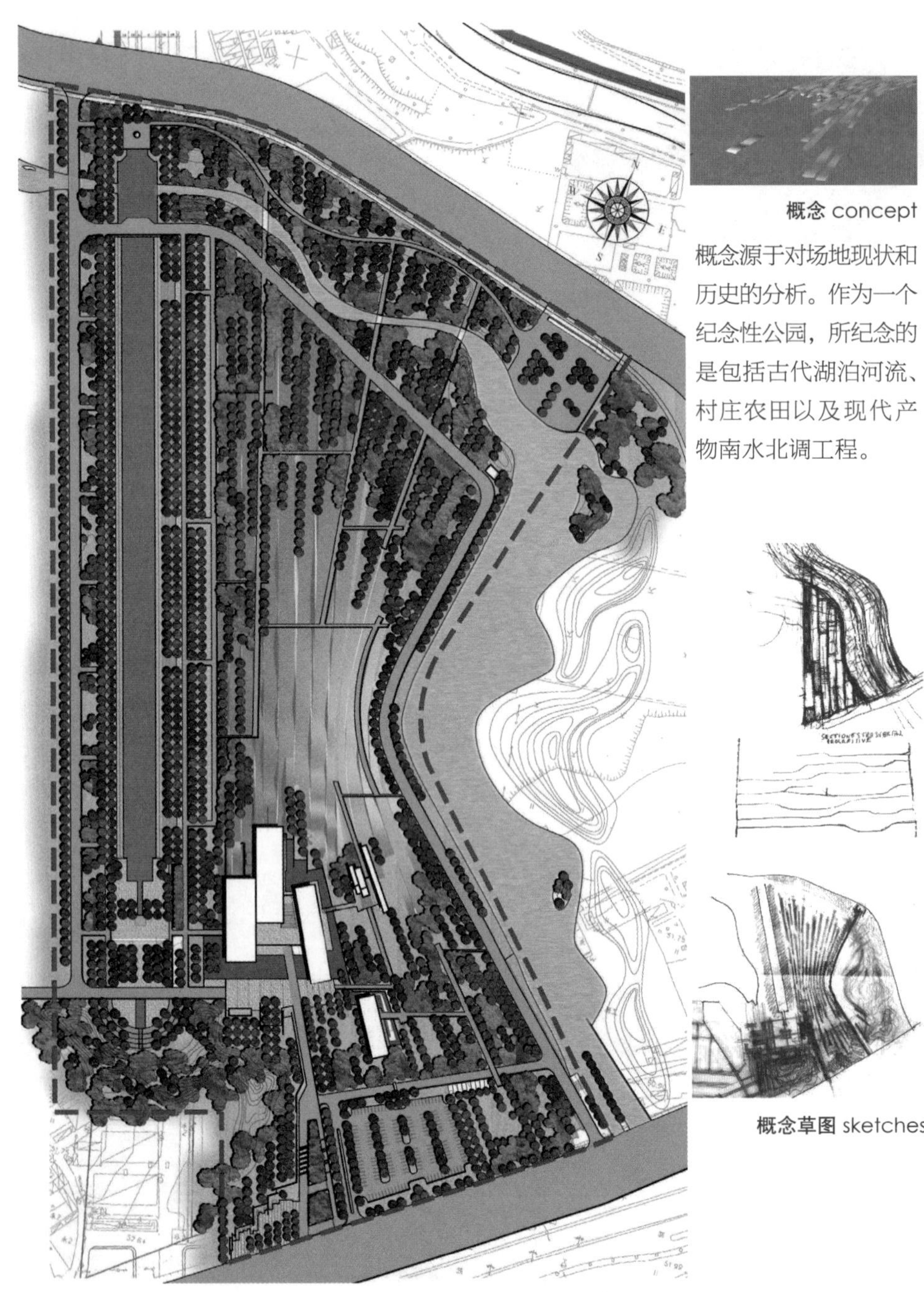

平面图 plan

概念 concept

概念源于对场地现状和历史的分析。作为一个纪念性公园，所纪念的是包括古代湖泊河流、村庄农田以及现代产物南水北调工程。

形式 form

这个压缩的形式来源于对现状因素的考虑。包括两条轴线和火车站的关系。

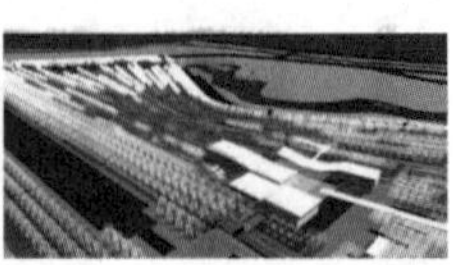

大地艺术 land art

创造新的地形的同时会产生新的景观和新的体验。通过种植添加色彩，通过交通系统增加人的活动空间。

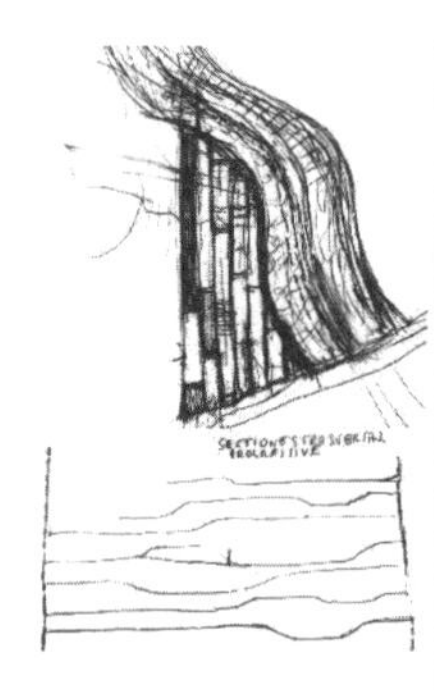

概念草图 sketches

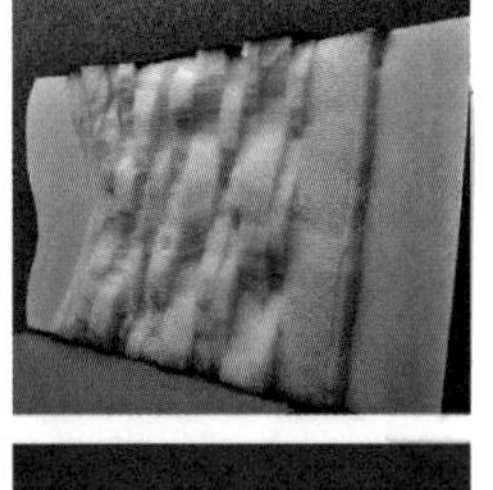

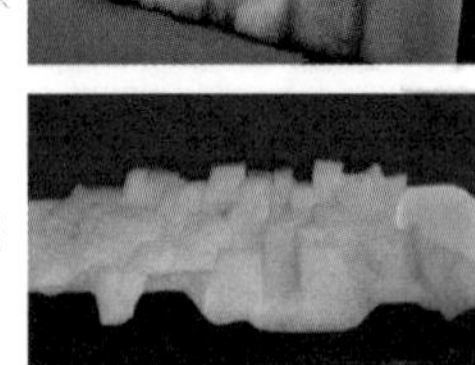

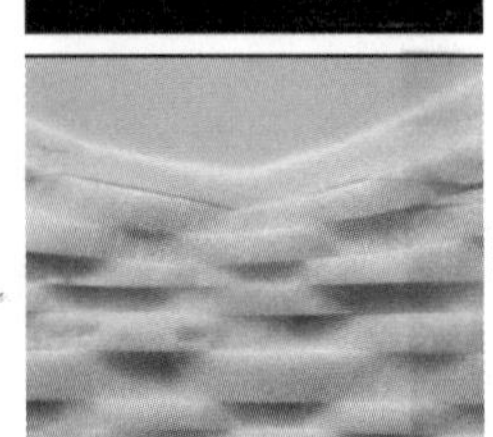

概念模型 models

景观结构分层 structure

竖向分析

Vertical Analysis

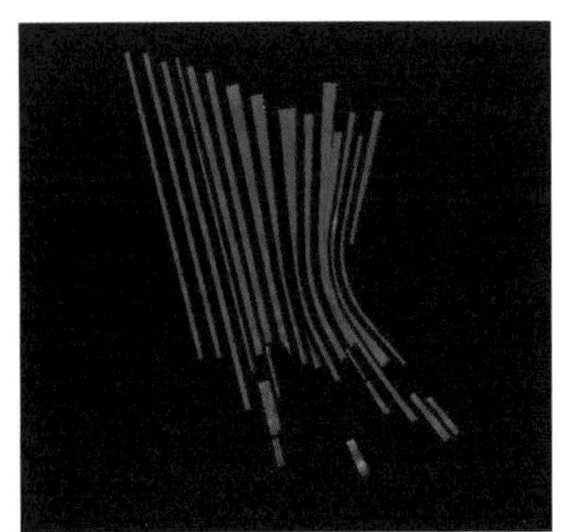

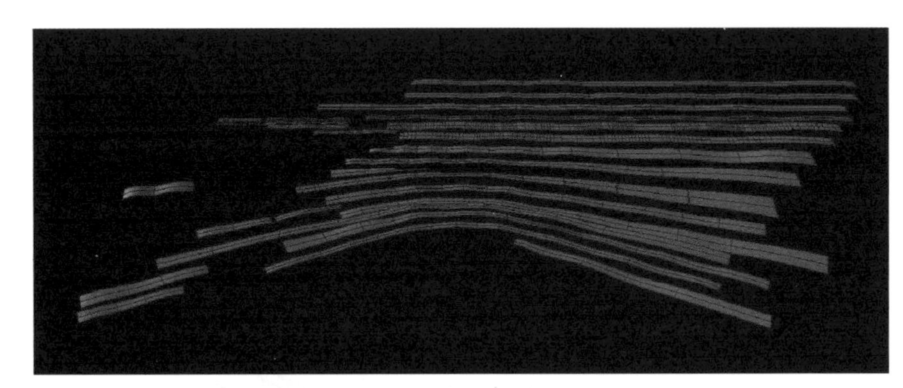

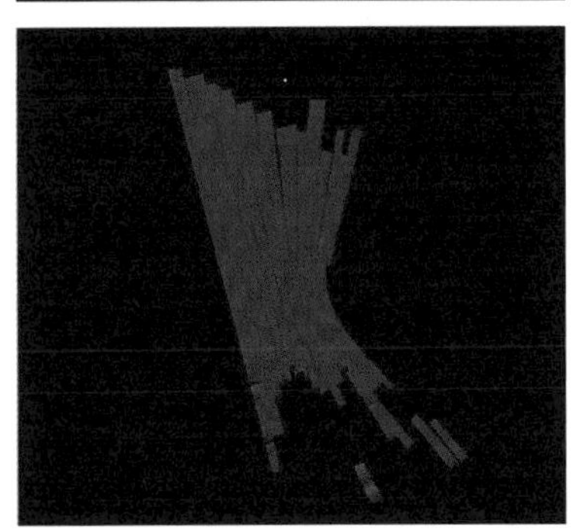

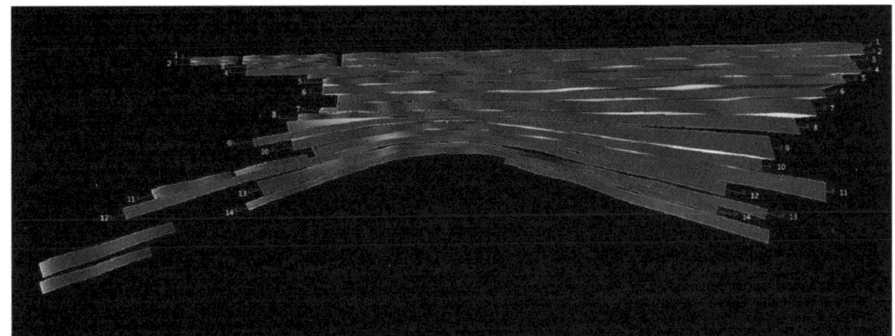

“大地艺术”是由若干个波浪形条带组成，通过对每个条带的剖面分析可以看到不同条带之间产生不同的空间。当我们建立东西、南北向的交通联系的时候，就产生了不同的道路剖面。

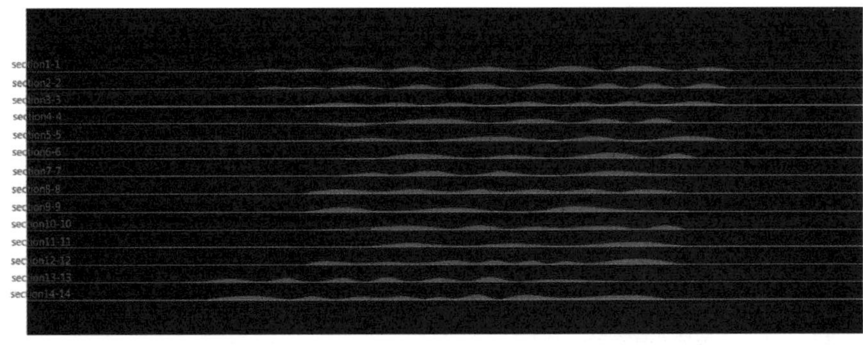

设计不同类型的道路连接南水北调水渠和主路，试图与大地艺术建立一个有趣的联系。按照与地形的关系，道路分成三种形式，高于地形、在地形上和穿越地形过。

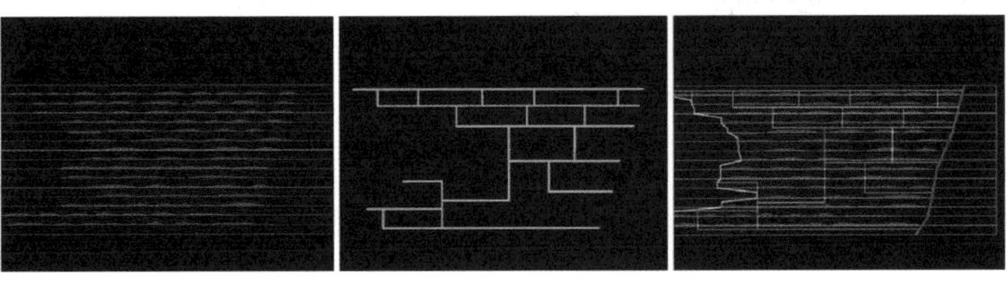

道路分析
Road Analysis

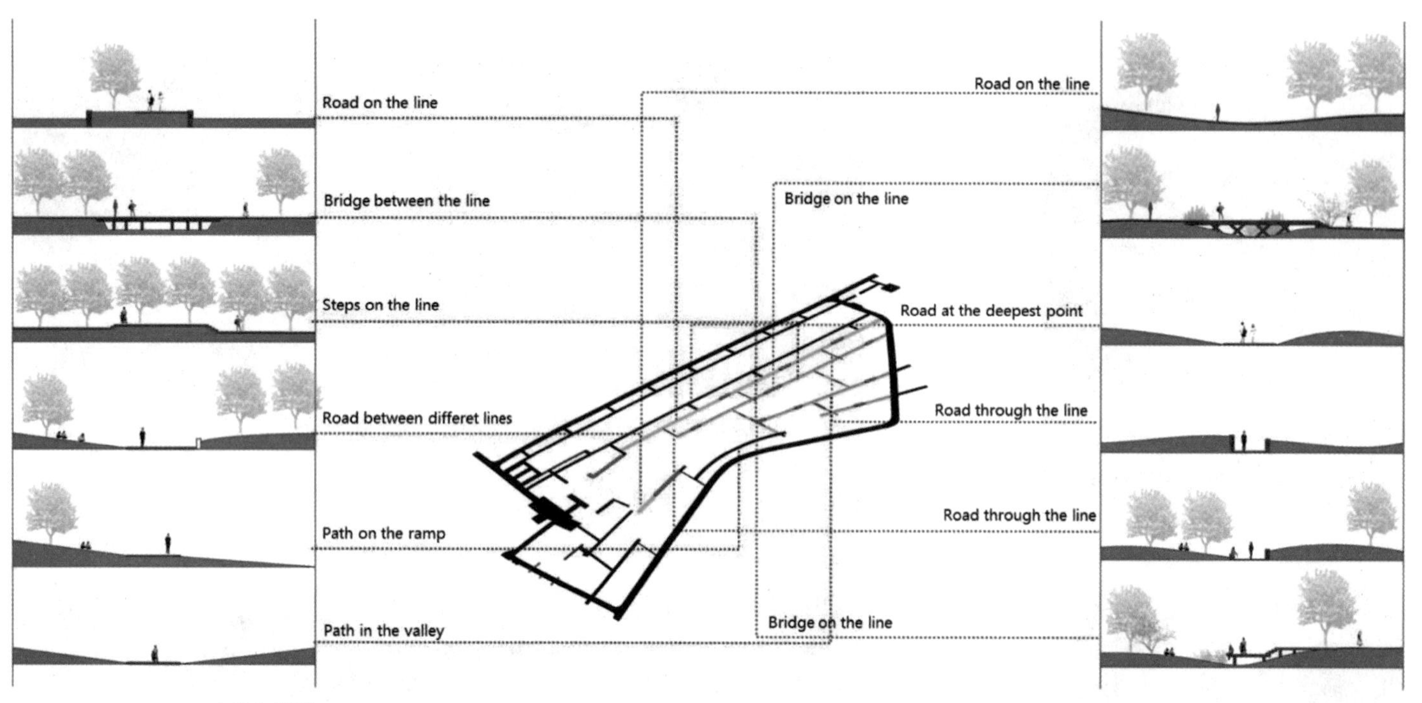

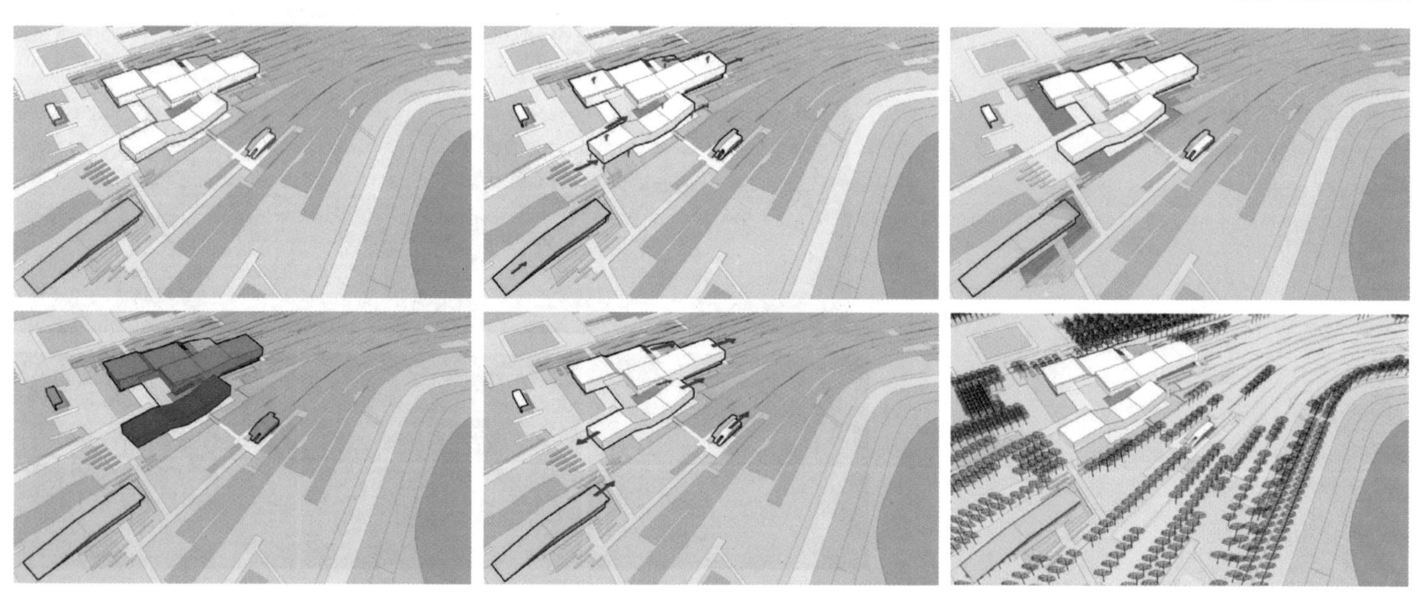

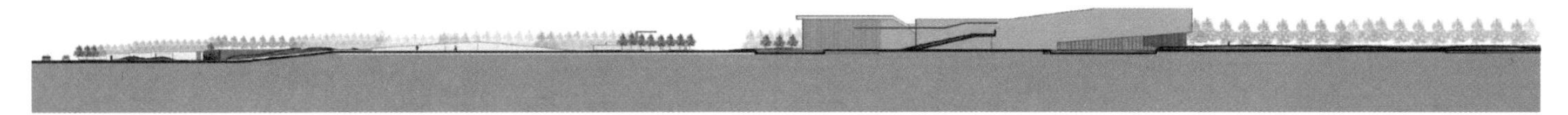

剖面图 section

透视图 perspective

融合

Fushion

赵婷婷　孙媛娜　武鑫
Zhao Tingting/Sun Yuanna/Wu Xin

城市、自然的手指相互交融，相互渗透。城市功能区包裹在自然之中，而自然功能区又赋予了一些必要的服务于城市功能的建筑，充分体现了我们之前的设想。

The city area, blending natural area each other, mutual penetration. Urban area peppered with natural landscapes. In natural area, new building to provide the necessary social services. Penetration and service facilities of the vision of the landscape interspersed distribution, fully embodies the concept of a handshake between city and nature.

01 理念构思 Concept

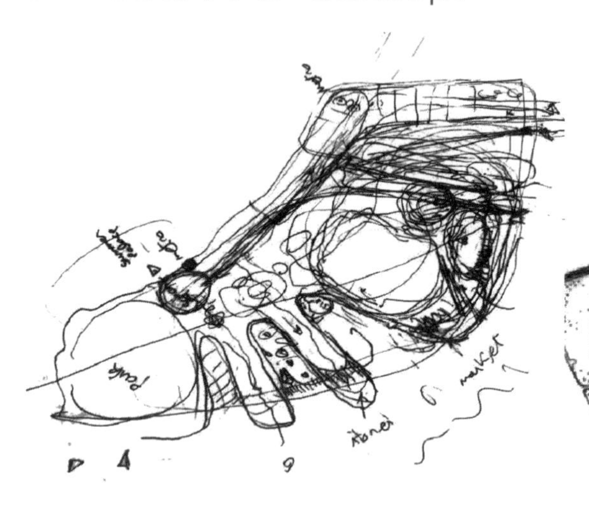

最终总平面图 final plan

概念规划 Concept

总体规划中的第一想法就是融合现在的城市边缘区域。这就是为什么提出将颐和园和城市连接在一起的原因。首先，这个方案努力将自然引入城市。

这种努力集中体现在我们认为重要的方面，住房、商业服务等。而且，我们方案中最重要的原则是保护调节池和原有的道路。保留的地方和整理出的地形组成了方案的骨架。

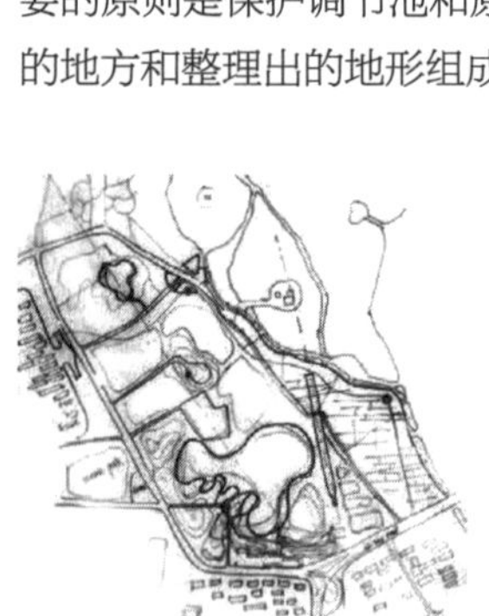

人流分析 spatial analysis

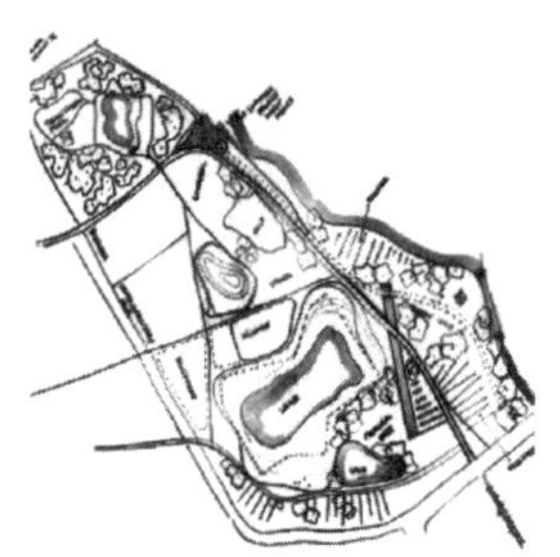

交通分析 traffic analysis

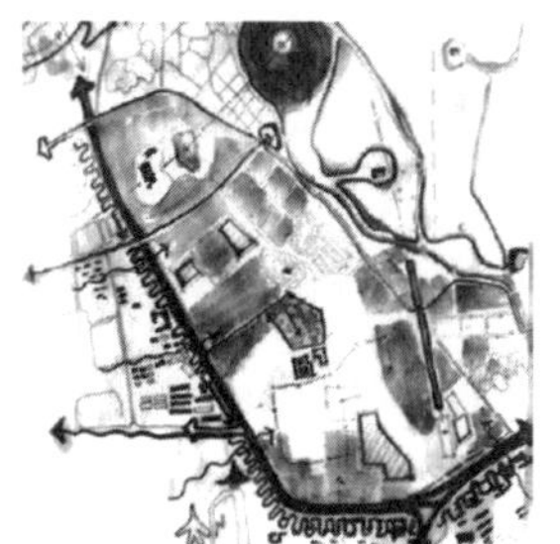

功能分区 function zone

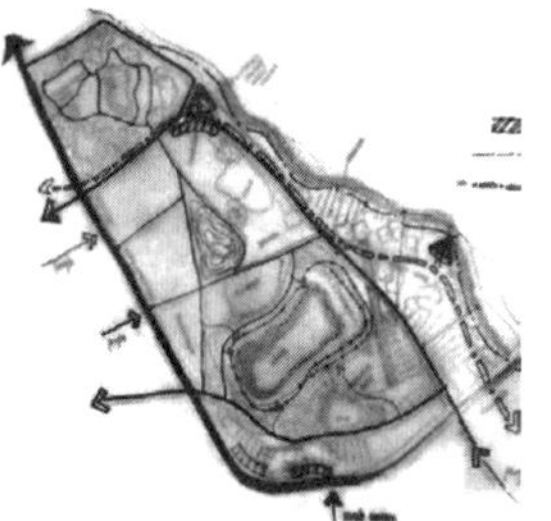

人流分析 spatial analysis

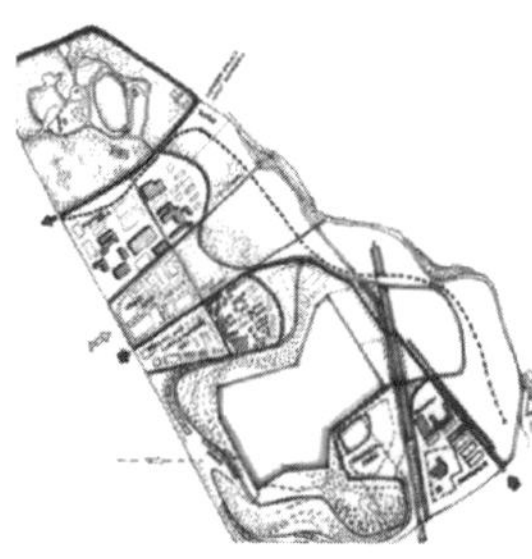

人流分析 spatial analysis

在最终方案中，城市、自然的功能相互交融，相互渗透。城市功能区包裹在自然之中，而自然功能区又赋予了一些必要的服务于城市功能的建筑，充分体现了我们之前的设想。

在北侧城市板块，改造原有建筑的立面及整合其功能，使之成为新型商务办公区。

主要服务中心，建立与之配套的社区服务、少年宫、幼儿园等建筑。沿北坞村路，建立大型超市，既满足原有社区人群需要，也可服务于北坞村路西侧社区的居民。

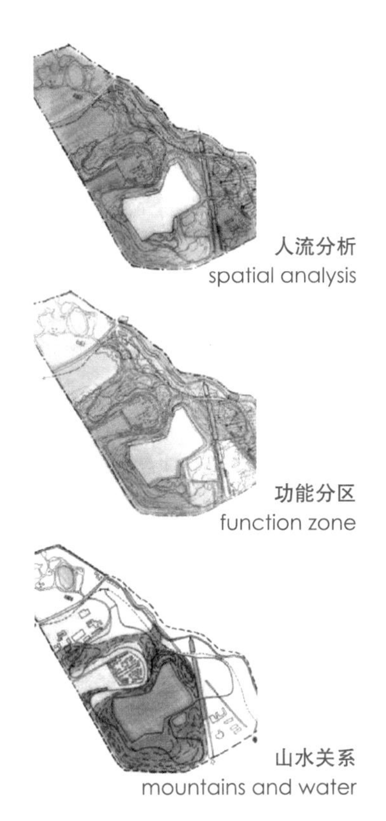

人流分析 spatial analysis

功能分区 function zone

山水关系 mountains and water

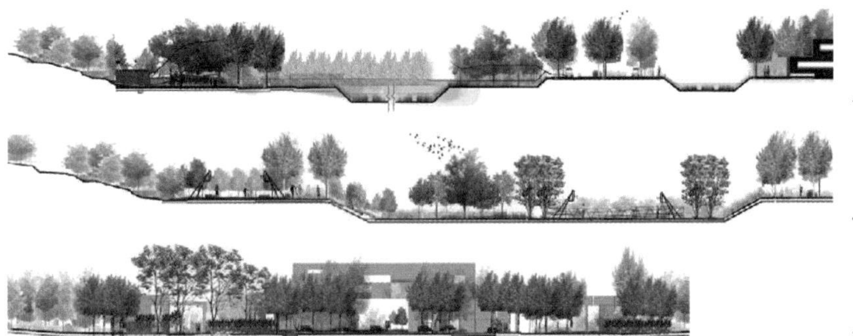

剖面图 A section A

体育公园剖面 section B

社区服务中心前广场剖面 section C

02 总体设计 General Design

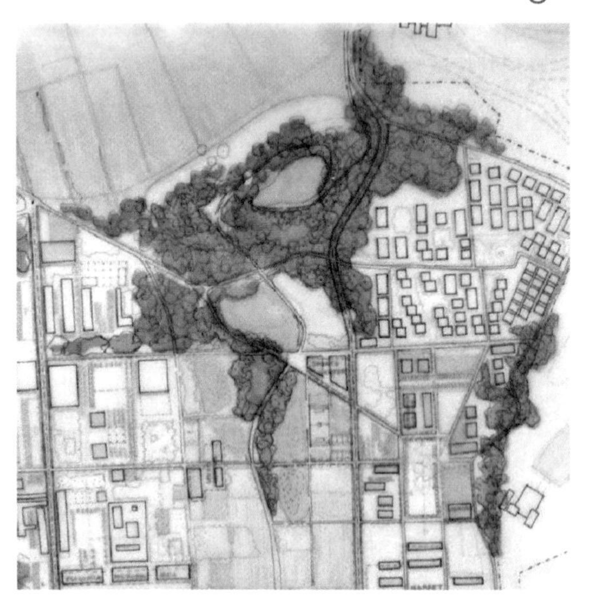

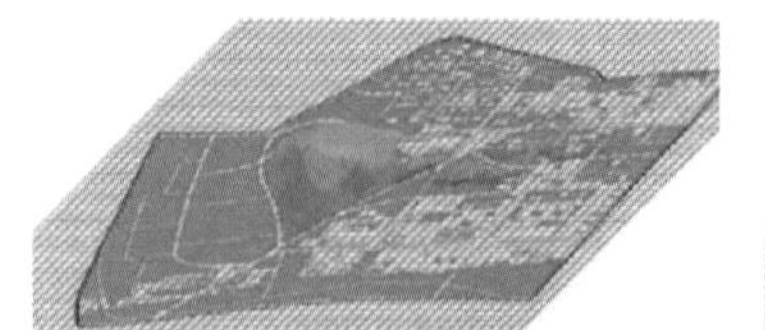

地形分析 geography

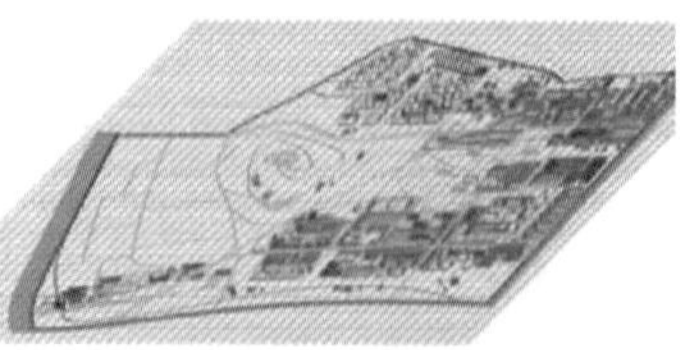

功能分区 function zone

道路分析 road analysis

建筑分析 building analysis

植物分析 plantings

在最终方案中，城市板块和自然板块交叉融合。城市区域始终融合着自然风景。在自然区域，新的建筑物提供了必要的社会服务设施。视线的流畅和服务条件的结合最终完整的诠释了自然结合城市的融合概念。

功能分区 function zone

03 详细设计 Detailed Design

最终平面图 plan

由于我们追求的是更加亲和、自然的公园环境。所以此区域内没有明显的入口标志。游人将通过园路与广场的引导，自然地进入园区。由于此站点是最靠近颐和园西门的轨道交通站点。这里将承担去往颐和园、北坞公园、玉东公园以及本园区大量游人的集散需求。分设在轨道两侧的集散空间将有利于分散不同目的的游人。

在这一区域内我们设计了一个同时具有公园管理功能、商业服务功能以及展示功能的小型展览馆。

建筑的一层局部架空，形成建筑的灰空间。在为游人提供展示、休憩等功能的同时，可以将建筑环境与自然环境相连通，通过这种方式进一步体现“融合”的主题。

展览馆建设成为采用中国传统建筑材料但风格更为简约的建筑。配以富有中国传统园林特色的景观植物进行点缀，从而体现一种古典文化与现代文化相融合的氛围。

稻田中流动的博物馆

Floating Museum in the Field

解陈娟　马双枝　李宏丽　慕晓东
Xie Chenjuan/Ma Shuangzhi/Li Hongli/Mu Xiaodong

以连绵起伏的地形塑造了有别于城市公园的山谷景观，这里更接近于自然。场地内的原南水北调水渠、泄洪渠均传递着场地的内在信息。通过对原南水北调水渠纪念性改造及对泄洪渠的艺术利用，使其与场地的稻田景观相融合。通过在其中建造实体博物馆而实现对场地记忆的展现。

The vally landscape using the rolling terrain shape makes the site different from the city park, here is more close to the nature. In the field, original South-to-north water diversion canals, release flood waters canal were telling the internal information. Based on the Memorial reconstruction of the original South-to-north water diversion canals and the art utilization of the release flood waters canal, make these form integration with the paddy field landscape. By the construction of entity Museum to display space memory.

01 场地现状 Site Analysis

总体介绍 Introduction

设计地块位于颐和园西南角，过去曾经是京西稻田所在地，后有当地村民居住于此，现在大部分居所已拆迁，仅存留一处居住区。场地关系复杂多样，西侧与城市相接，东侧与颐和园相邻，同时规划中的南水北调工程、有轨电车、现状的水渠、排洪渠、苗圃、厂房、市场以及北侧的北坞公园等均交织于此。

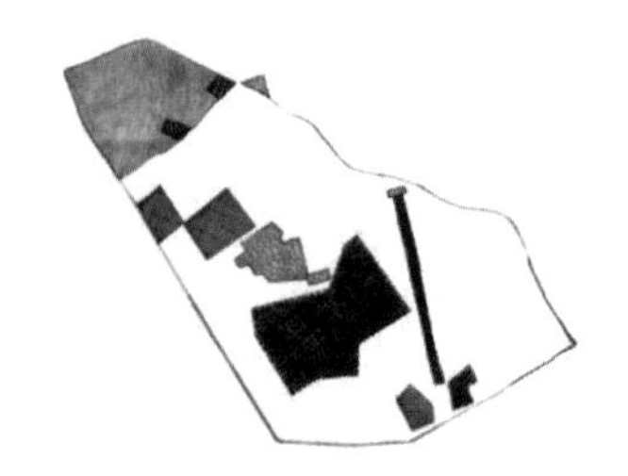

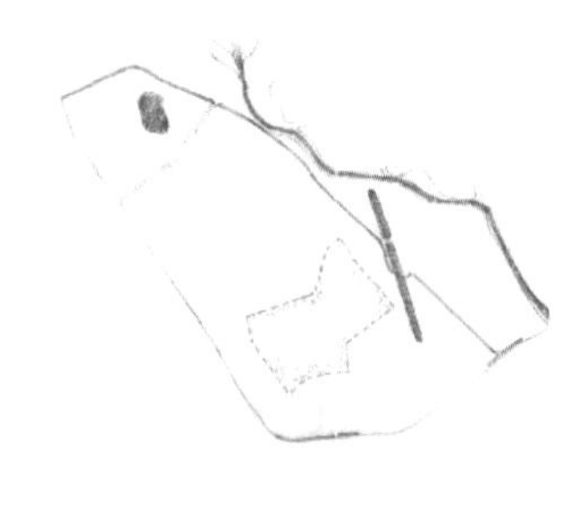

场地现状分析 site analysis

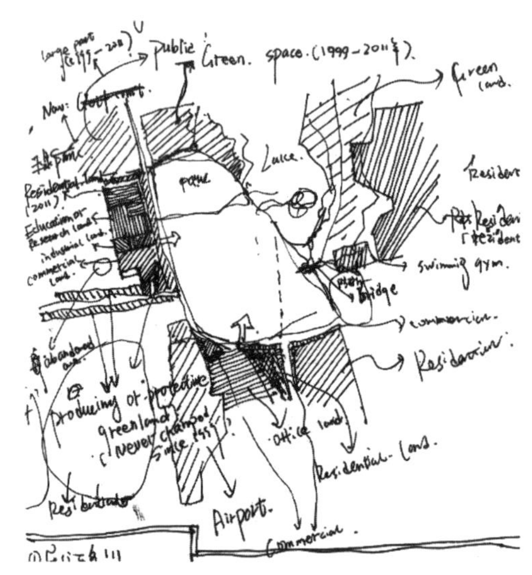

场地适宜性分析 suitability analysis

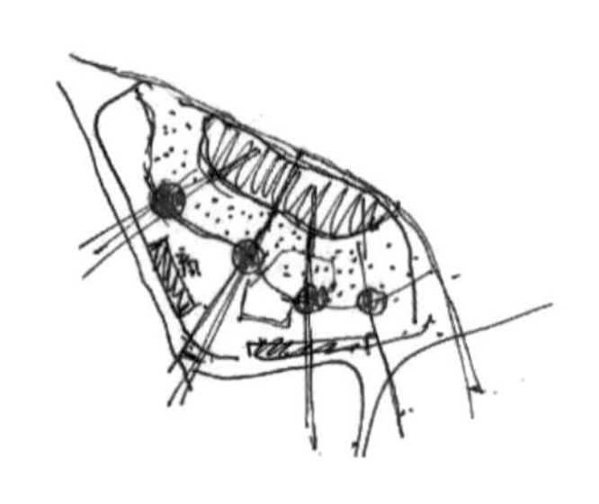

“城市—自然”概念图示 concept of city-nature

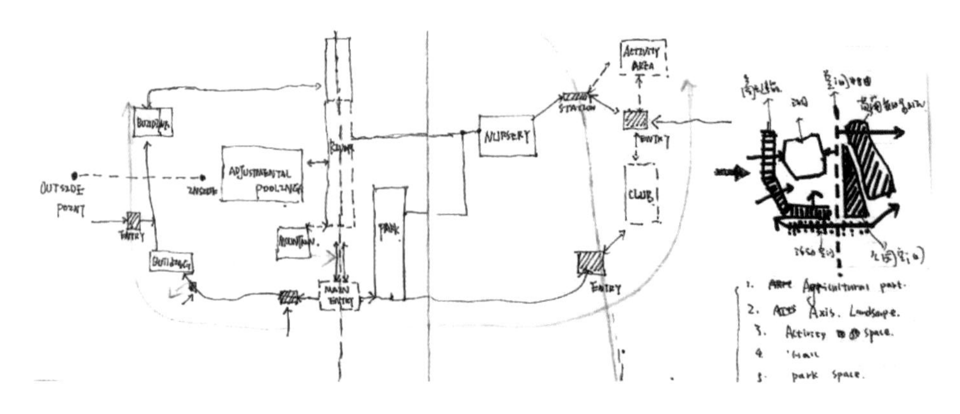

“城市—自然”过渡关系分析 analysis of relationship

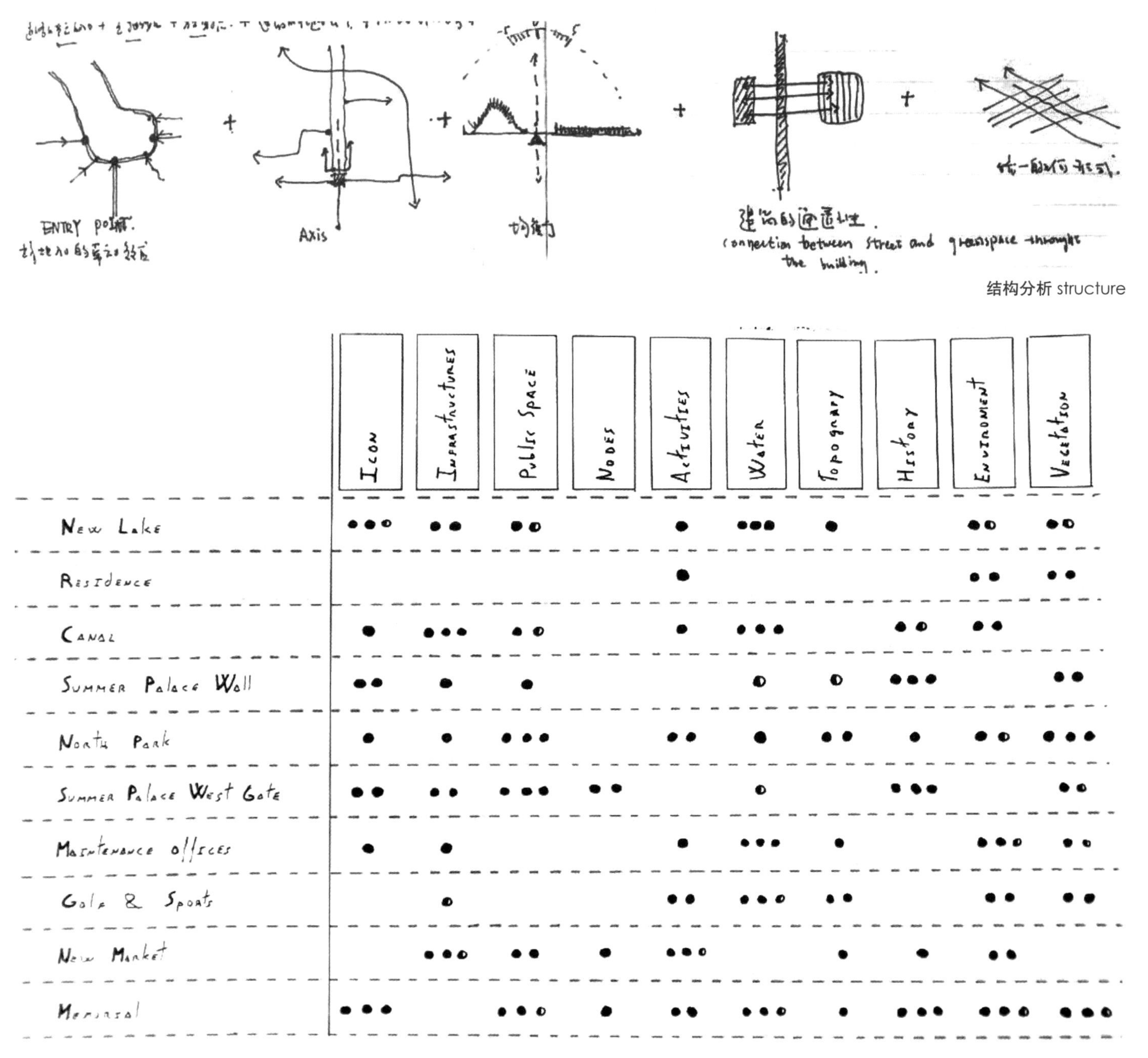

结构分析 structure

功能需求分析 function demand

30 公顷方案设计

Design for the 30 Hectares

以连绵起伏的地形塑造了有别于城市公园的山谷景观，这里更接近于自然。场地内的原南水北调水渠、泄洪渠均传递着场地的内在信息。通过对原南水北调水渠纪念性改造及对泄洪渠的艺术利用，使其与场地的稻田景观相融。通过在其中建造实体博物馆而实现对场地记忆的展现。

在原南水北调水渠的南段设置博物馆，位于公园主要入口处地形下，既提供了入口视线的障景又充分与水渠相结合，是水渠纪念性的展示窗口，内部提供有关南水北调的历史展览；水渠两侧设计游人可亲近的亲水平台，在游玩中感受历史。位于稻田内部的泄洪渠，利用埋于地下的端头处进行放大，结合其水面设置另一博物馆，此博物馆用于展现此处的京西稻田的演变历史以及与颐和园相关的历史内容。

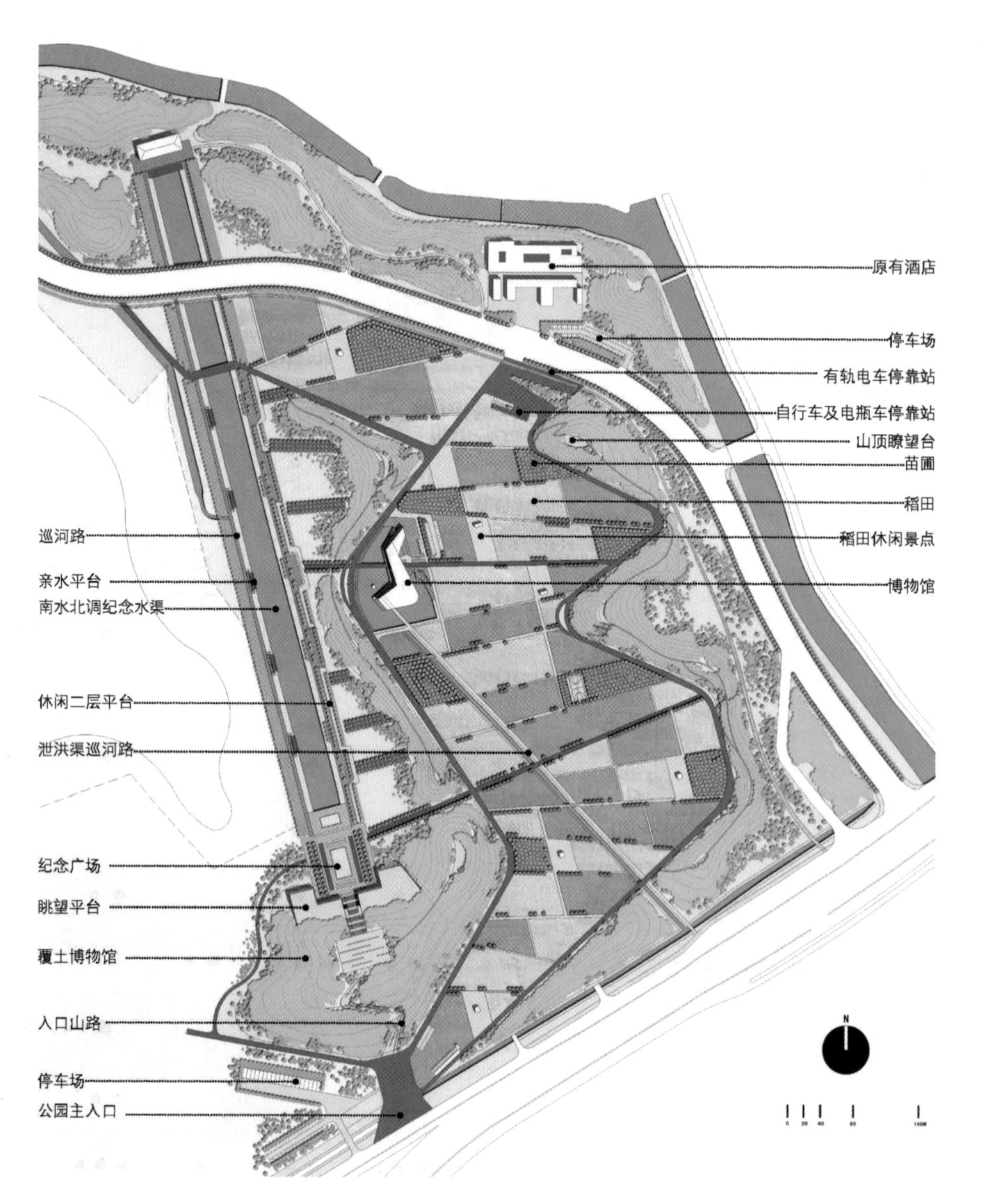

总平面图 plan

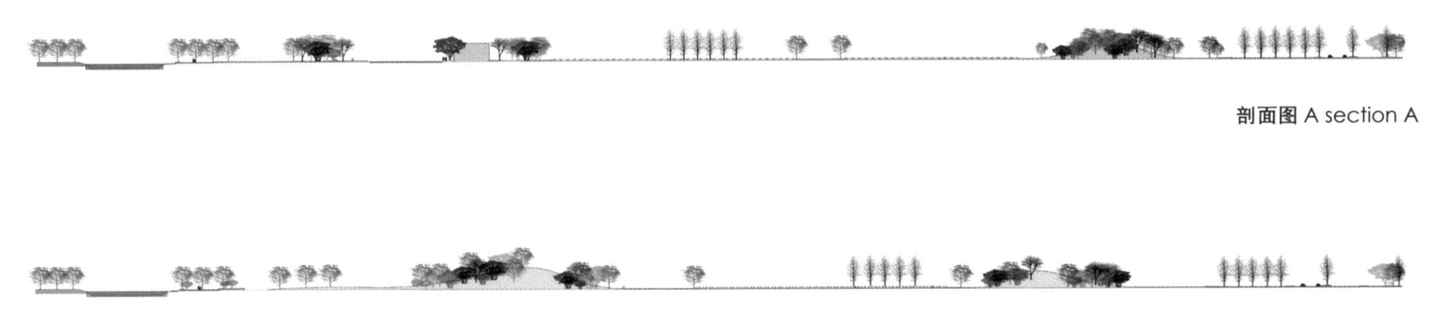

剖面图 A section A

剖面图 B section B

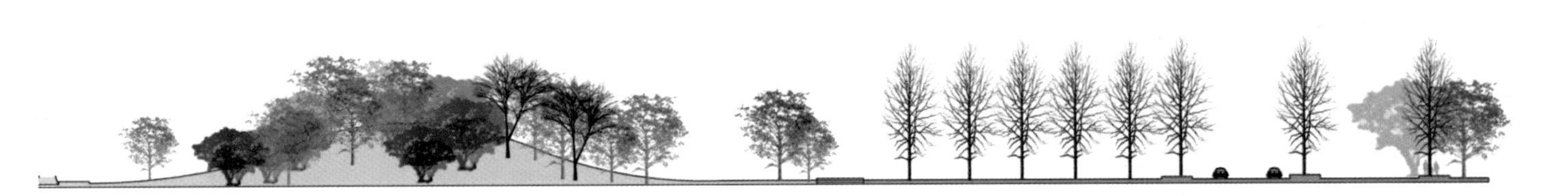

放大局部剖面 enlarged section

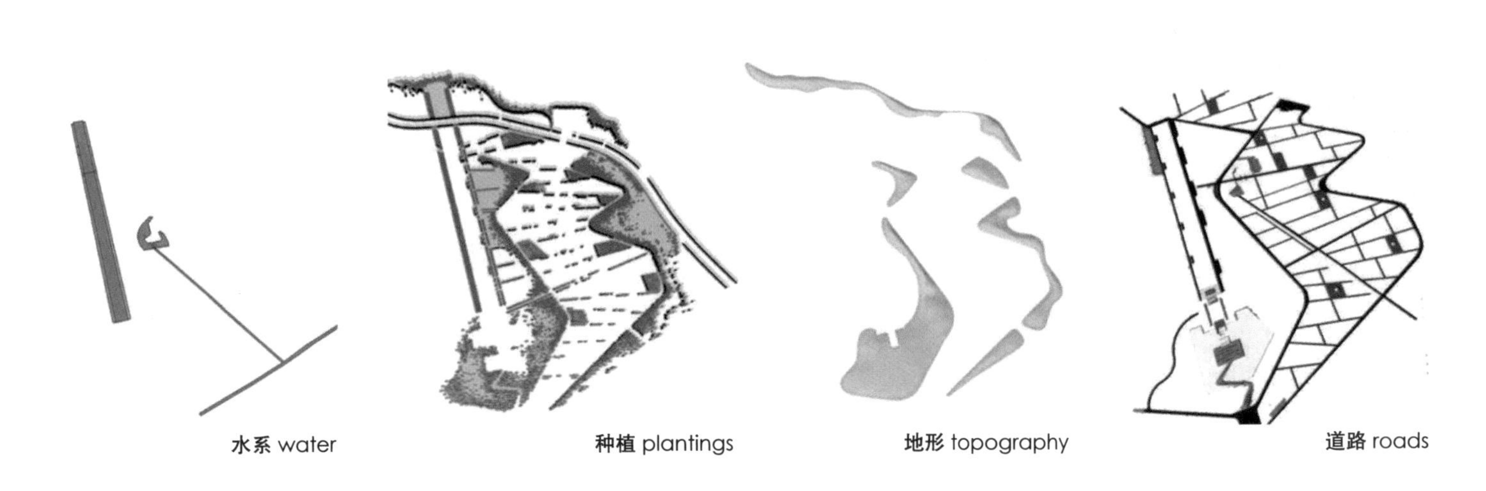

水系 water　　种植 plantings　　地形 topography　　道路 roads

03 模型展示 Design Model

此处西邻原南水北调水渠，东与颐和园有规划城市道路相隔，两座山体恰居于此，形成了静谧的山谷稻田的景观。通过城市道路、山体道路、内部的稻田小路使场地的游线得以贯穿，并形成不同的视域景观。

模型平面 model plan

模型剖面 section

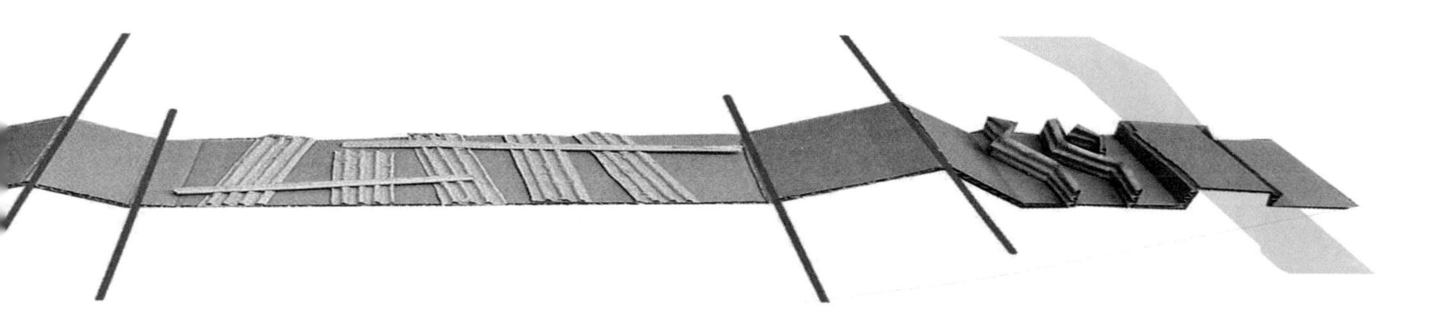

水体及道路示意 water and road

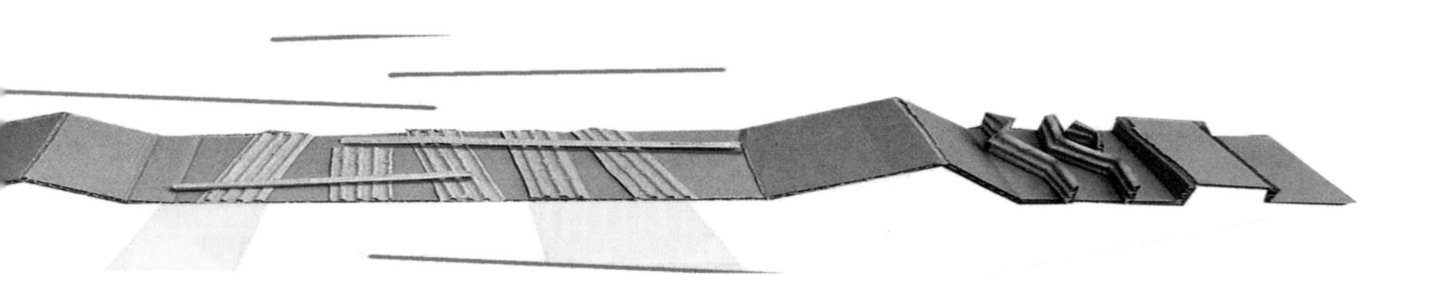

稻田及小径示意 paddy fields and path

5 公顷方案设计 Design for the 5 Hectares

这是结合泄洪渠的水体而设计的博物馆及其周围的稻田景观，此博物馆与稻田中的主要东西向道路和水渠的巡河堤相结合，是重要的景观汇集点。游览于南北向的道路的人们会发现位于稻田景观中的各种功能的活动场所，如采摘亭、体验园、篮球场等，让人们能在闲暇的午后或游览完颐和园后有个别样的休息停留空间。

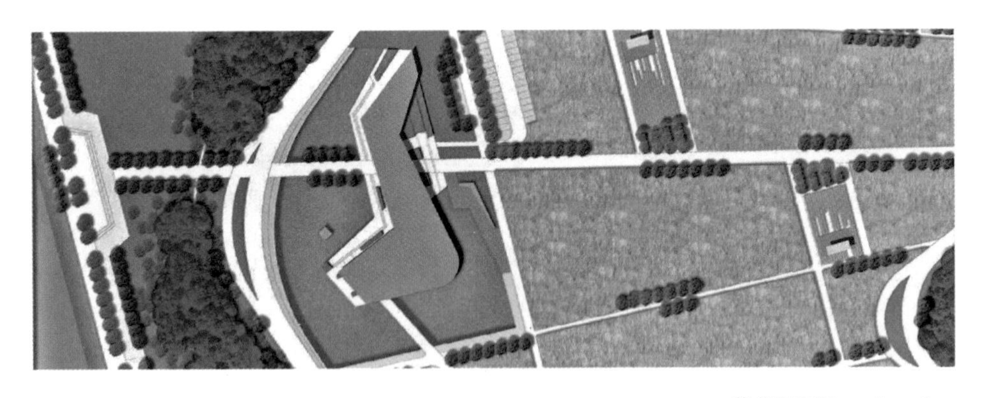

节点平面图 nodes plan

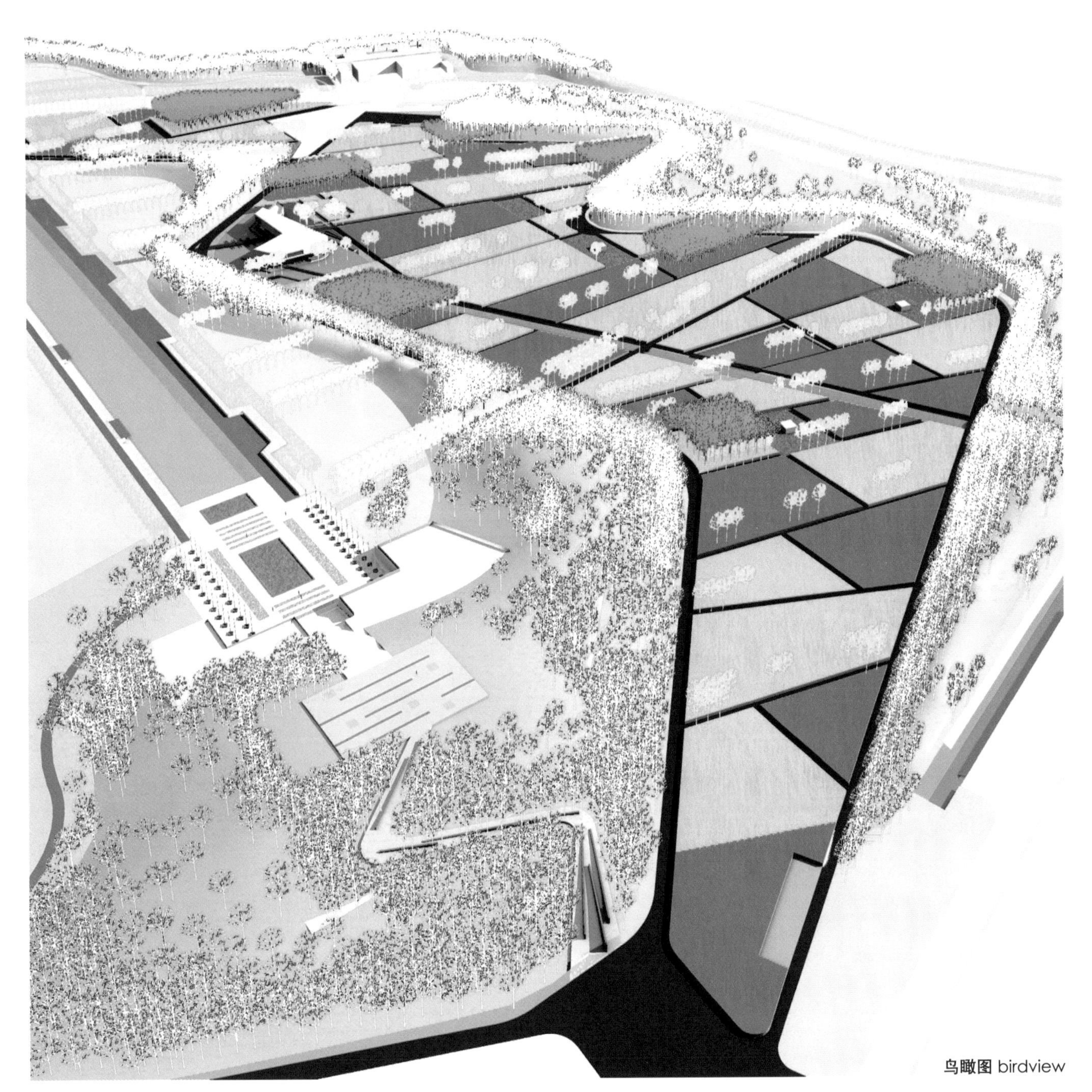

鸟瞰图 birdview

叠变

Translay

杨永亮　赵维佳　刘畅
Yang Yongliang/Zhao Weijia/Liu Chang

南水北调工程的建设可以说很大程度地破坏了地表径流，使雨水发生乱流。为解决这一问题，我们在调节池和明渠周围重新梳理了地形，引入雨水花园的概念，并做了一些亲水设计，在雨季和旱季都可游可赏。

The transfer project actually destory the runoff of the site. In order to solve the problem, we created some new terrains and introduced the concept of Rain Garden and designed some hydrophilic facilities to make it both usable Inrainy and dry seasons.

01 理念构思 Concept

场地解读 Site Analysis

场地地理位置较为特殊，地处“三山五园”历史景区。北面世界文化遗产颐和园，西望玉泉山玲珑塔，南面和东面是农民拆迁移居的小区。场地自身承载着大量历史信息，曾是最富盛名的京西稻田，是古今调水文化的历史见证。

一级叠变 first level

二级叠变 second level

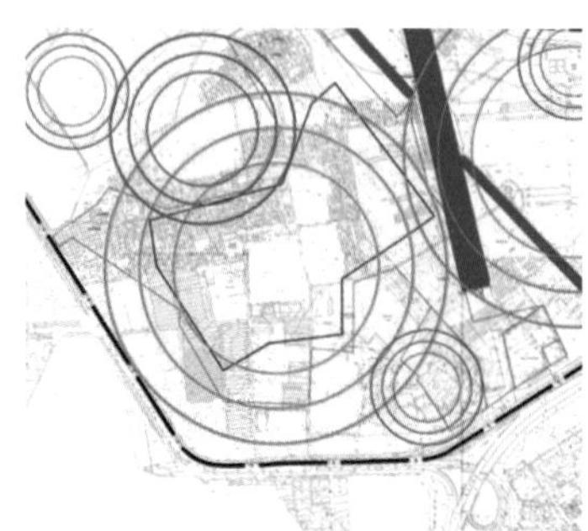

三级叠变 third level

概念解读 Concept

总体规划部分，概念的衍生源于场地自身属性。可以说空间上，场地是古典园林与现代城市以及特殊工程地段的叠加；时间上是覆盖了相当多的历史信息。所以说这是一个交叠性、辐射性很强的场地。设计者希望在设计中体现“叠”这种特性。并通过“叠”的手段使场地发生一些特殊的改变——这就是变的部分。“叠”从空间属性入手，“变”则从解决问题入手。南水北调引水渠及调节池建成后，将极大程度上影响地表径流，设计通过重新梳理地形，叠加一些小水系，组织地表径流。

最终总平面图 plan

纪念广场设计 Design of the Memorial Square

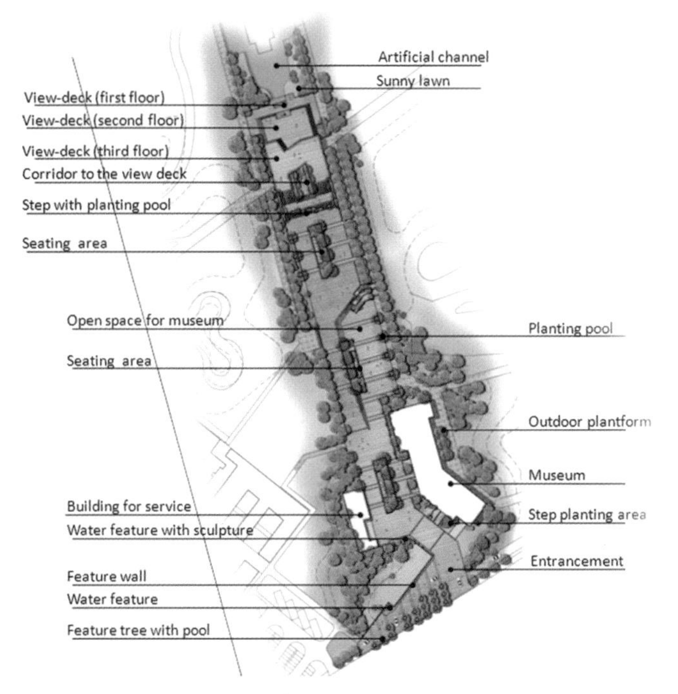

纪念广场平面图 plan of memorial square

剖面 A section A

剖面 B section B

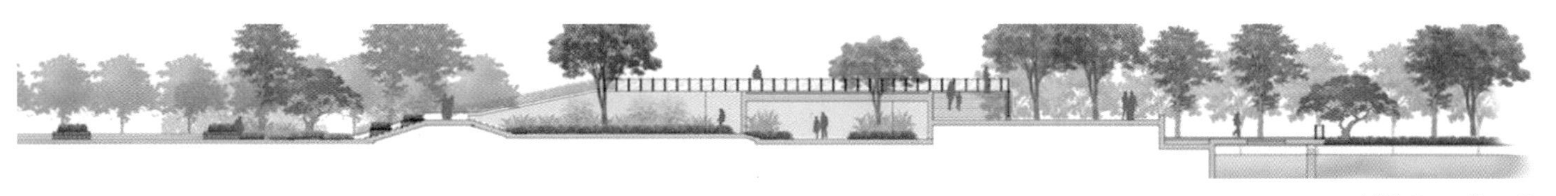

剖面 C section C

编织公园

Weaving Park

童牧
Tong Mu

我们希望能将山水田这些元素“编织”进场地，不仅恢复该场地特有的景观，而且能将这景观与场地现有人们的需求组合在一起，做到人与自然的结合，才是这个场地的精神所在。

We hope to weave mountains, water, field into a characteristc landscape, and it will be integrate with the current situation. It is the spirit of site that man is an integral part of nature.

童牧

Tong Mu

01 场地现状 Site Analysis

概念解读 Concept

通过对场地的调研，我们注意到该场地的两个特点：

1. 场地位于“三山五园”景区内，紧邻颐和园，北面是玉泉山，西面能看到西山余脉，场地周围有良好的中国式的山水景观。

2. 该地区在历史上曾是皇家御用的稻田，在当时曾是著名的农业景观，被称为“京西稻田”。该地与皇家园林有着联系，但它有自己的历史与文化，形成了自己独特的景观：既有风景如画的山水景观，又有与生产相关的农业景观。但随着时间流逝，场地的变化翻天覆地，这个特点也已然消失。

我们希望能将山水田这些元素“编织”进场地，不仅恢复该场地特有的景观，而且能将这景观与场地现有人们的需求组合在一起，做到人与自然的结合，这才是这个场地的精神所在。

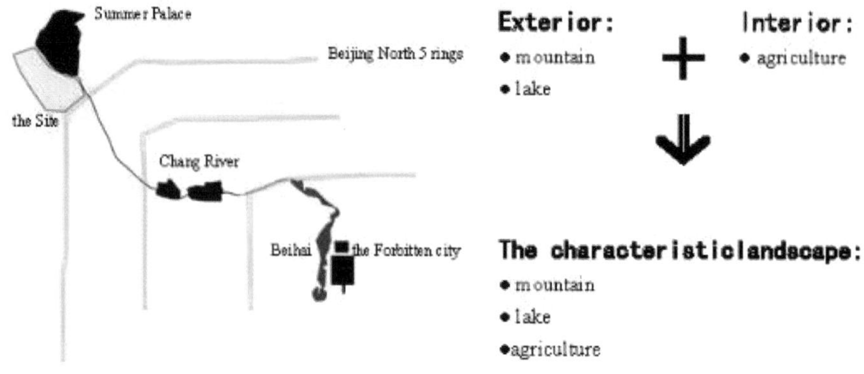

The site was associated with the Royal Garden.

It was different with the Royal Garden

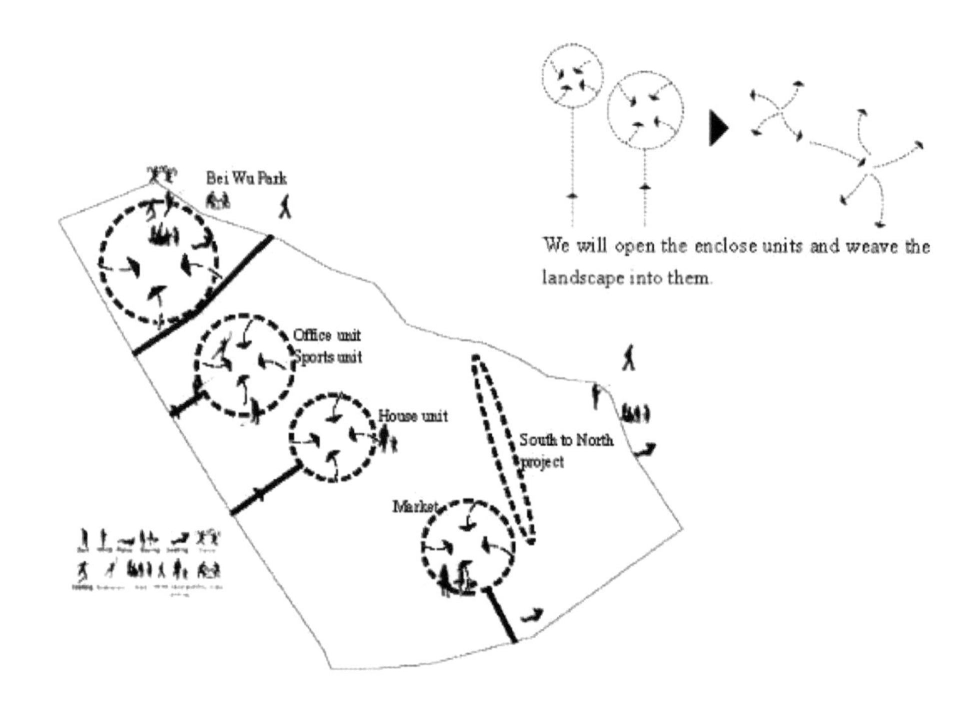

We will open the enclose units and weave the landscape into them.

There are many units for different activities now, but they are closed and lack of contact with each other.

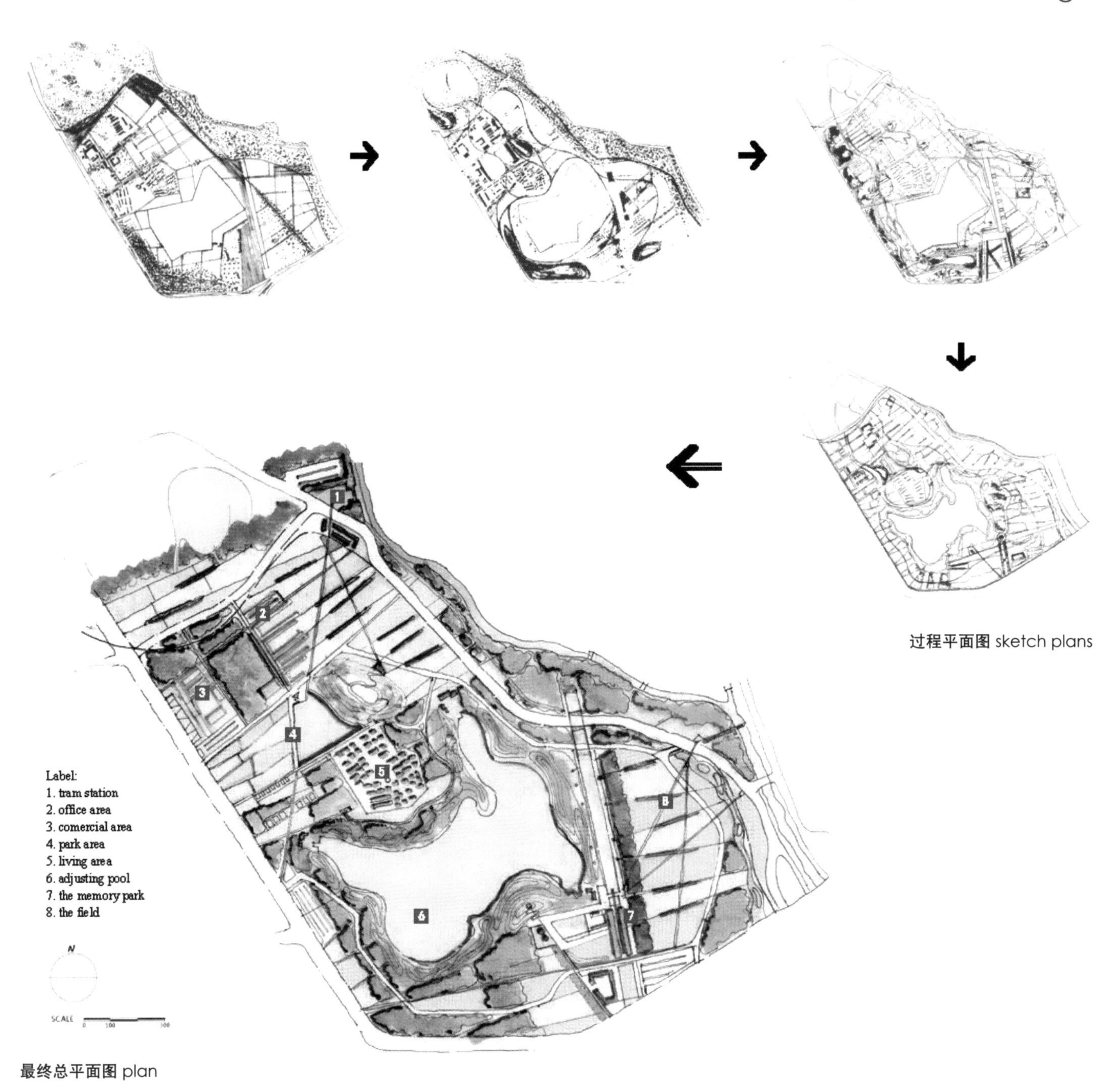

过程平面图 sketch plans

最终总平面图 plan

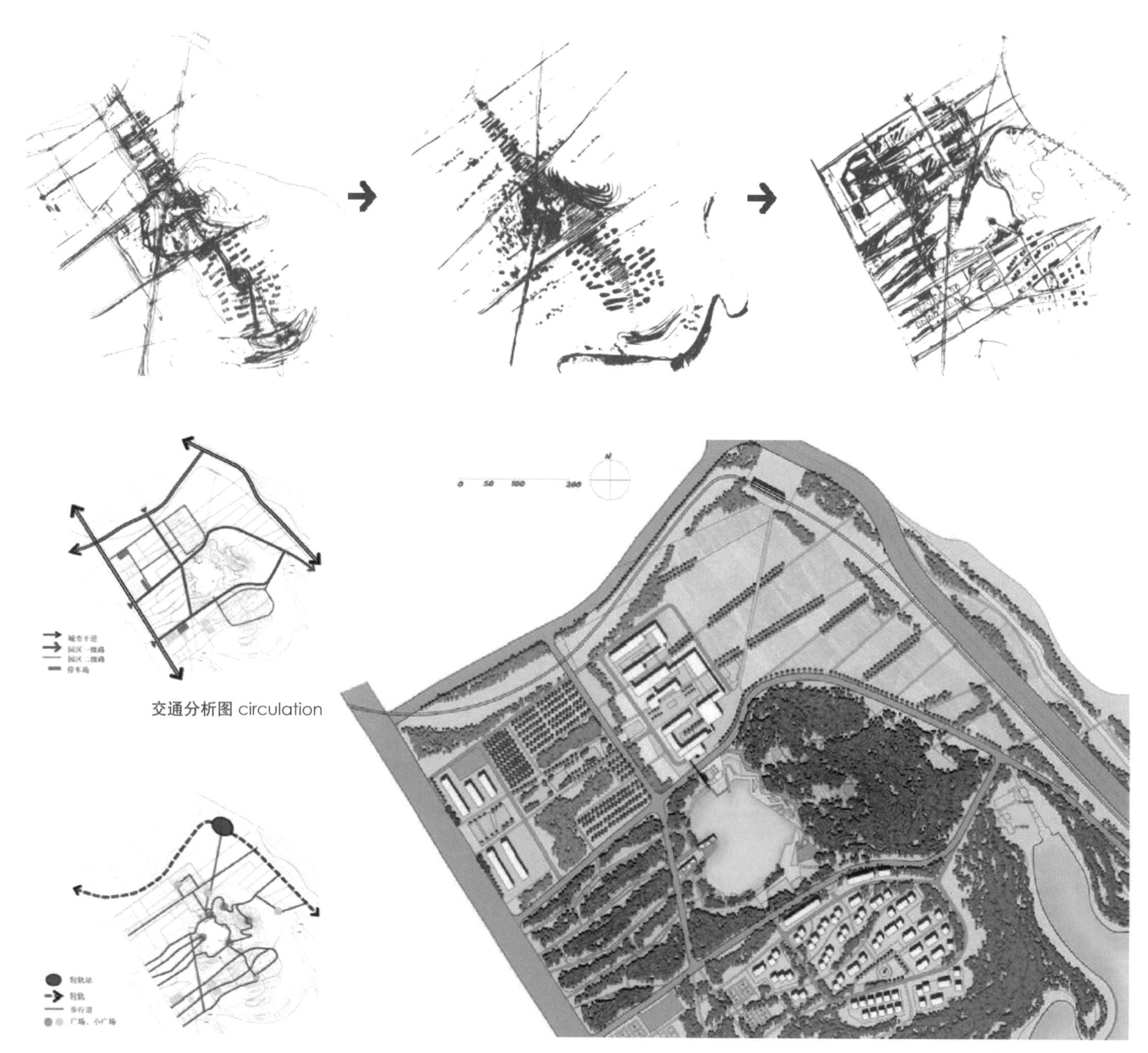

交通分析图 circulation

放大平面图 blow up plan

竖向设计 Topography

本场地除了现存的一座山外，高差变化非常小，场地十分空旷。因此我们利用山体、建筑与湖面的高差变化，创造了多层次的景观视线。同时，为了产生公园和城市之间的联系，通过塑造地形产生了东西向的引导，希望能够把部分城市人流引入公园中。

剖面 A section A

剖面 B section B

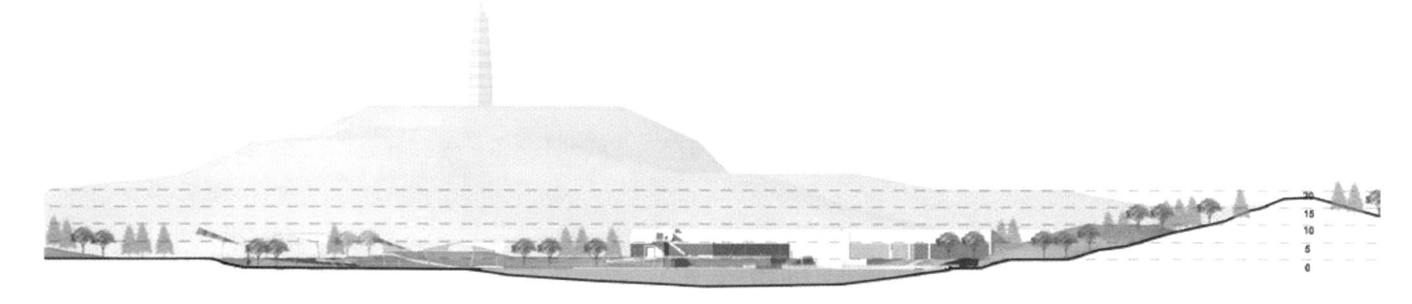

剖面 C section C

STUDIO PECKING

THU-AA 联合设计课程清华大学部分

2013.9 ~ 2013.12

组 1
Team 1

曹木　陈康韵
Cao Mu / Chan Hongwan

湖北黄石矿坑是历史遗留的尺度巨大的采矿遗迹，是人工作用于自然所产生的独特的空间。伴随着开采的停止，矿坑也成为鲜有人问津的地方。而此次对其景观化的设计，希望能够使矿坑再一次成为人工与自然对话的空间。

Huangshi pit in Hubei is a large-scaled mining relics left over from history. It is also a unique space produced by power of both human and nature. As the mining operation is stopped, the pit is going to be a negative place. This studio work, we want to use approaches of landscape design to revive the pit, transforming it into a place where combined artificial and natural power.

曹木
Cao Mu

陈康韵
Chan Hongwan

01 场地现状 Site Analysis

Meshing Principles

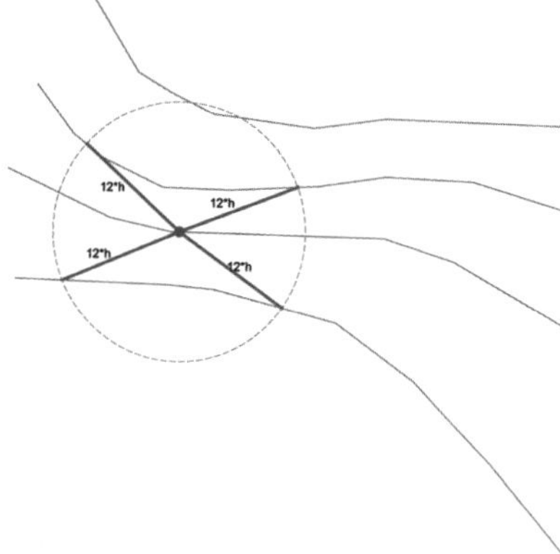

Agriculture as main activity:
accessibility on foot
topography

1:12 easy walking for pedestrians

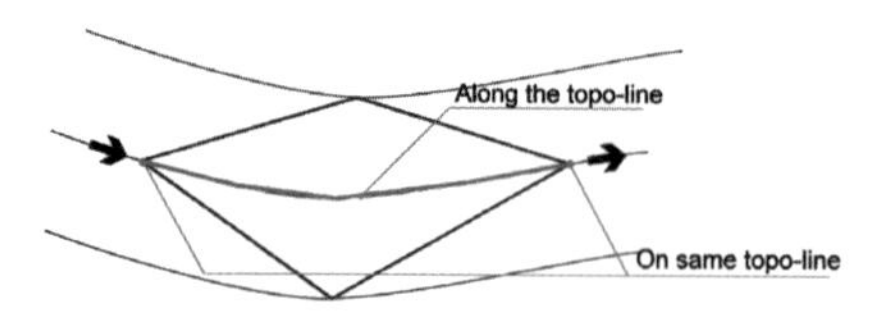

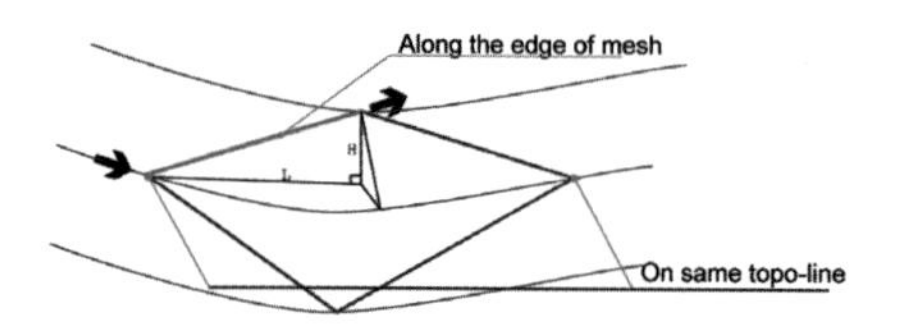

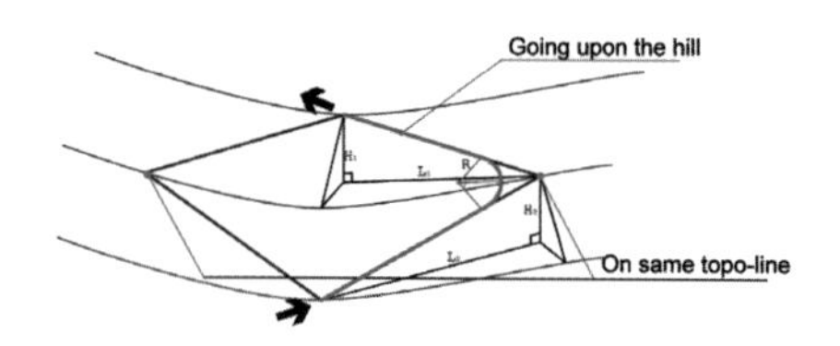

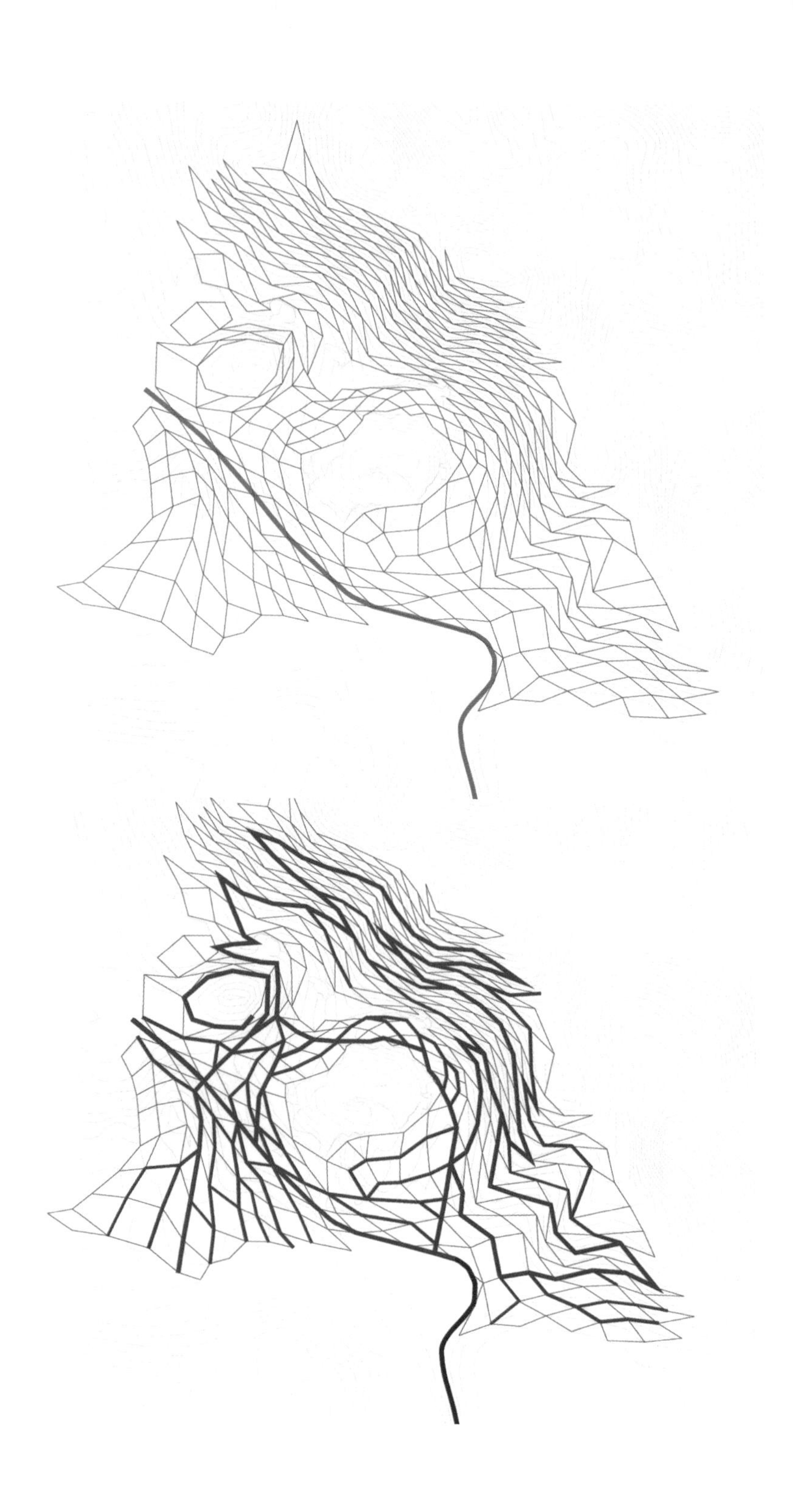

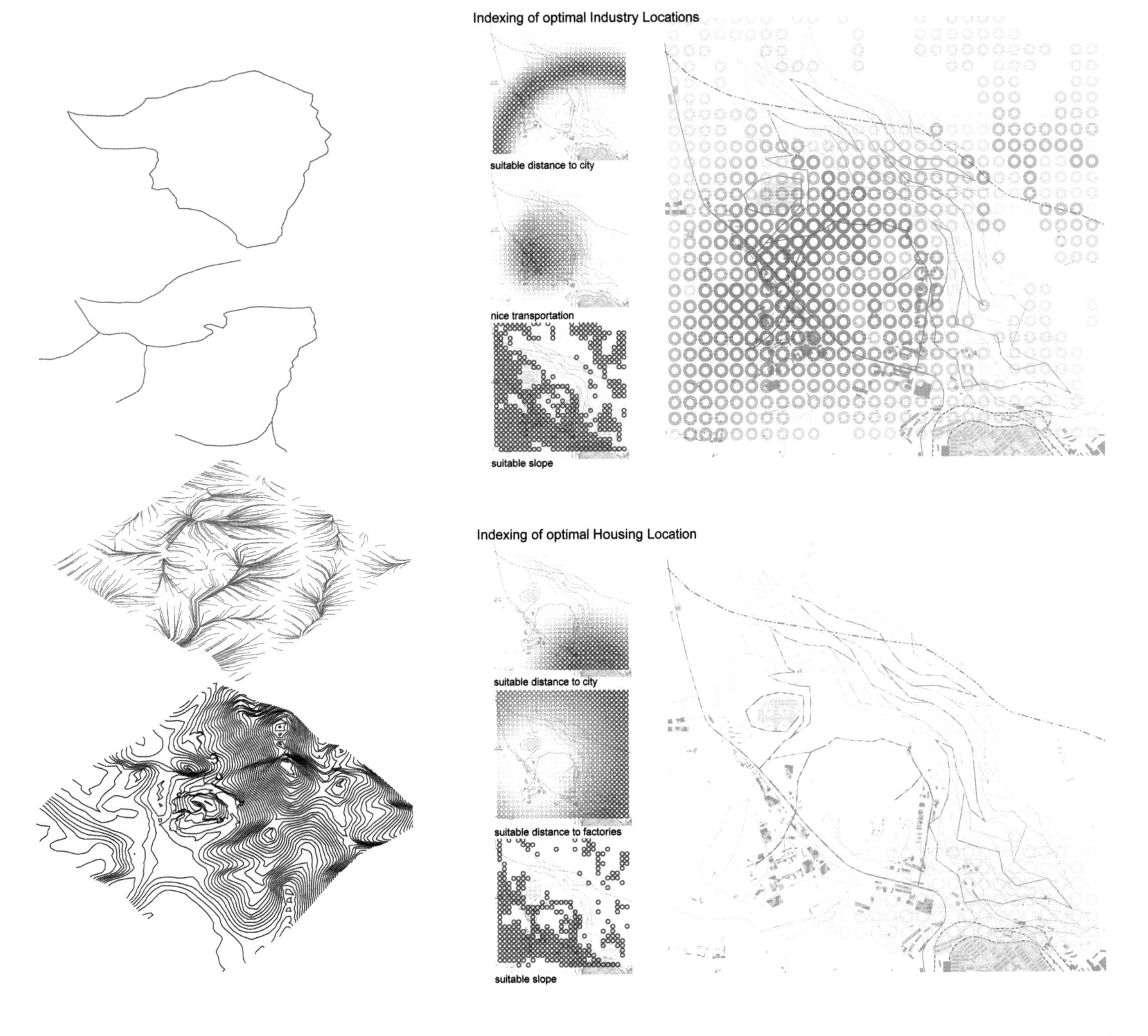
Indexing of optimal Industry Locations
suitable distance to city
nice transportation
suitable slope
Indexing of optimal Housing Location
suitable distance to city
suitable distance to factories
suitable slope

Subdivision principles

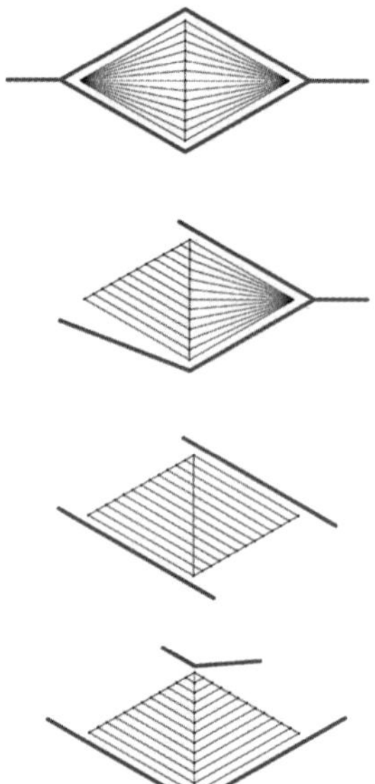

Prototype possibilities

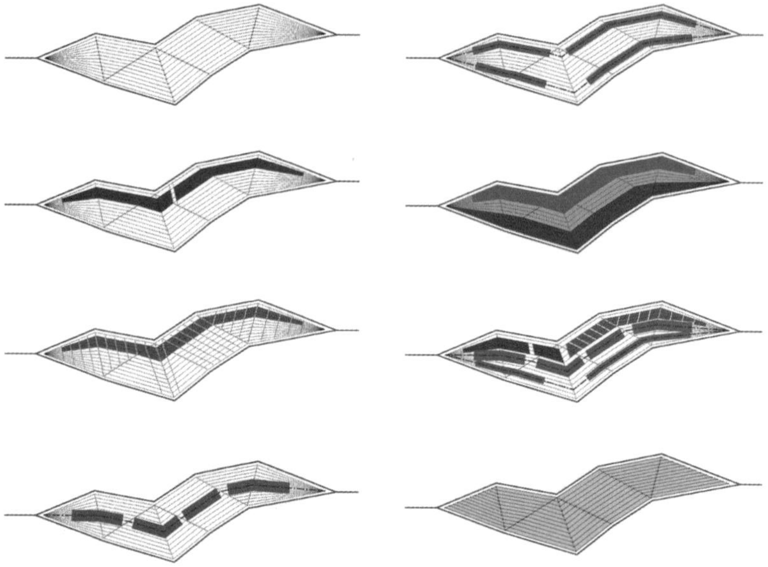

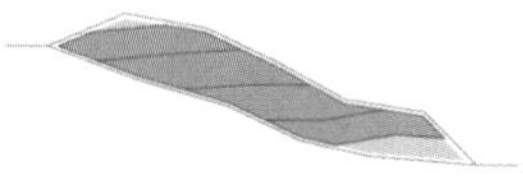

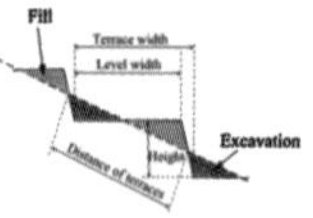

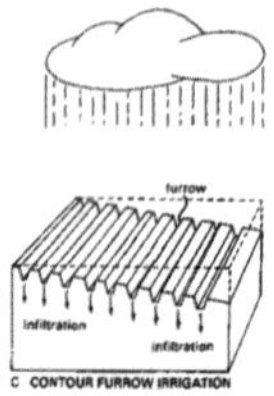

Artificial Wetland System

Agricultural Fields: Surface Flow Wetland
Residential Area: Subsurface Flow Wetland

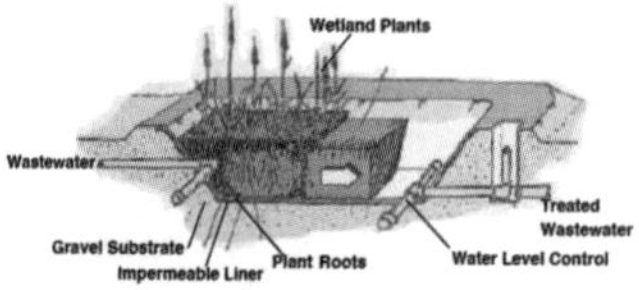

Wetland Plants
Wastewater
Treated Wastewater
Gravel Substrate
Impermeable Liner
Plant Roots
Water Level Control

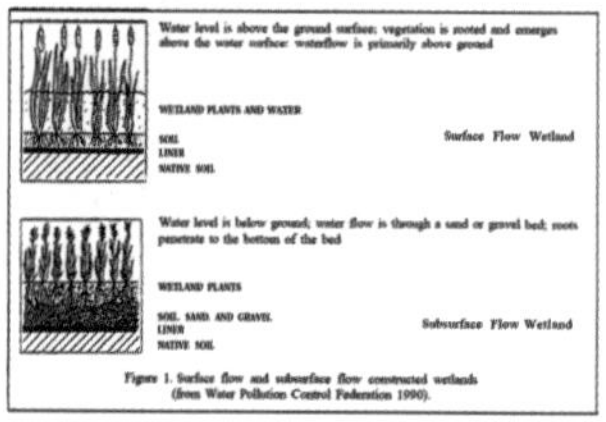

furrow
infiltration
infiltration
C CONTOUR FURROW IRRIGATION

Fill
Terrace width
Level width
Height
Excavation

Agricultural/Residential Wastewater
Rainwater
Sediment Basin
Artificial Wetland
Clean Water Pond
Irrigation canals
Furrow Irrigation

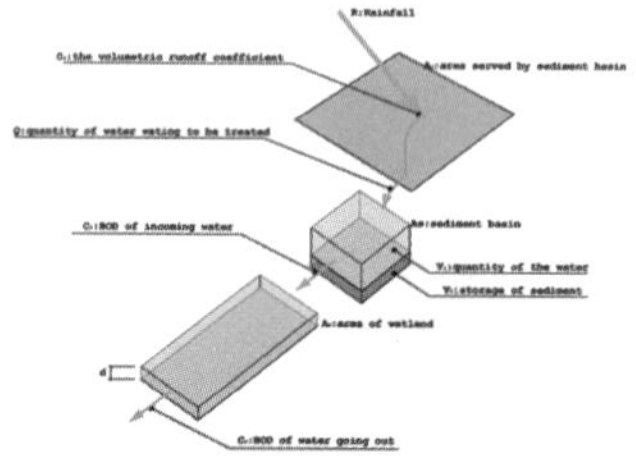

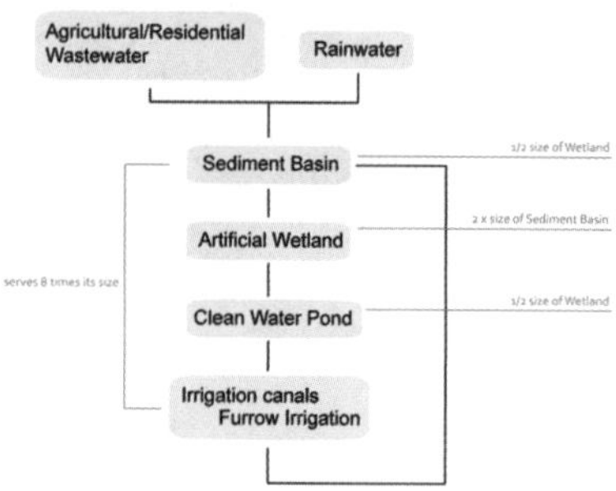

Indexing Water Catchment
For Wetland Locations

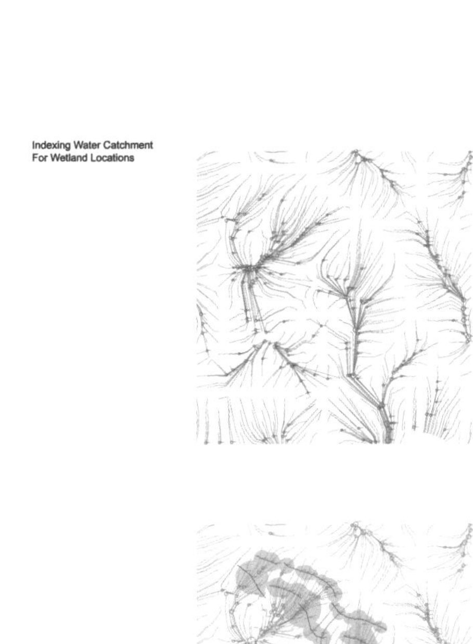

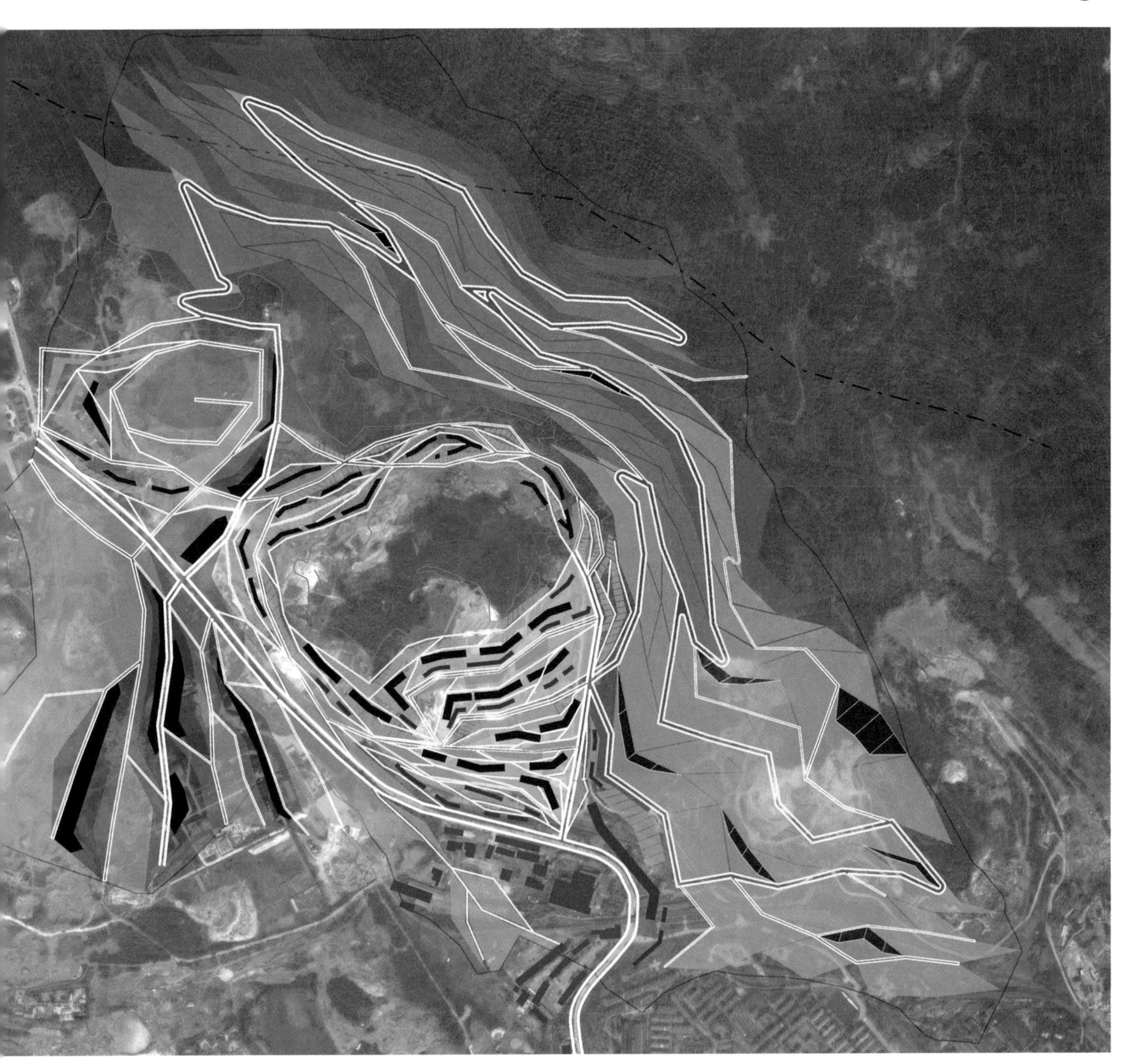

04 模型展示 Design Model

Prototype for Public Infrastructure

Shops
Kindergarten
Primary School
Sports Facilities
Workshop Spaces
Community Centers
Cafés
Restaurants

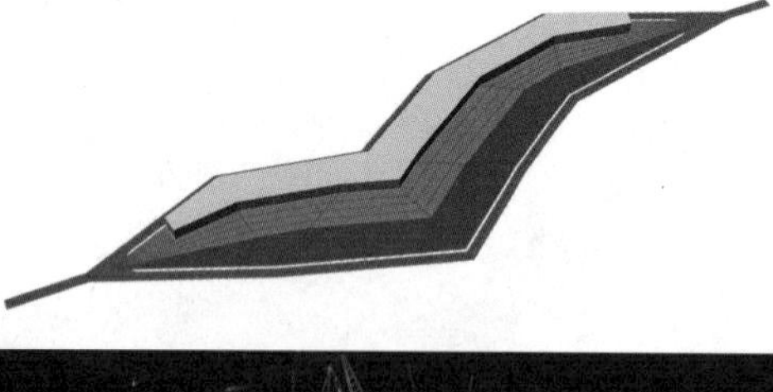

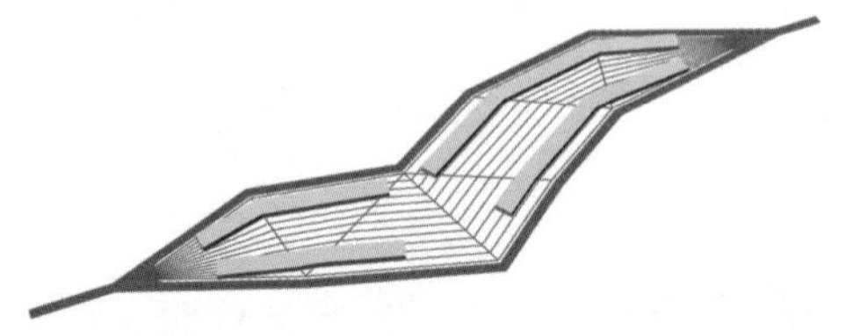

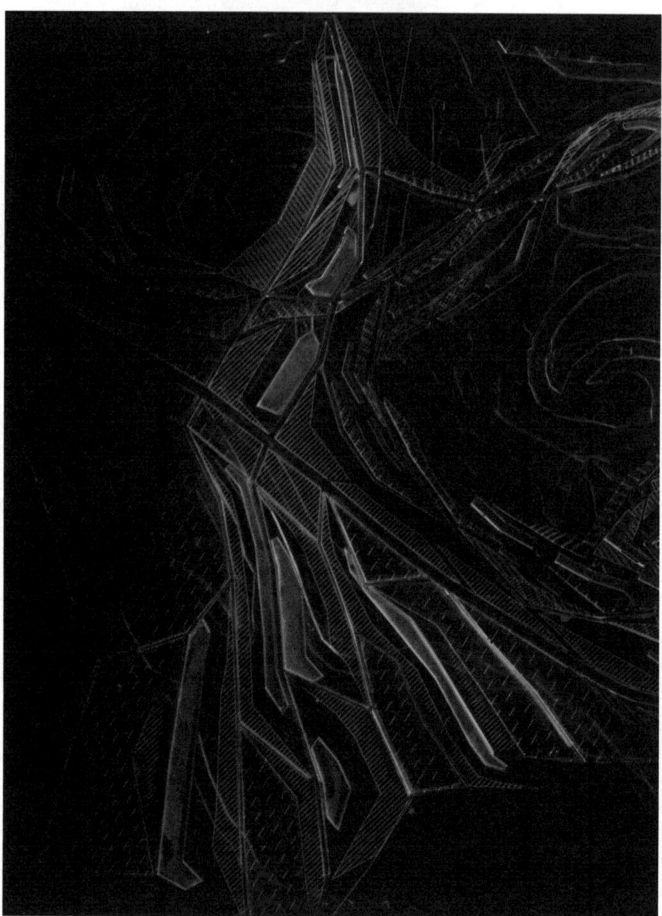

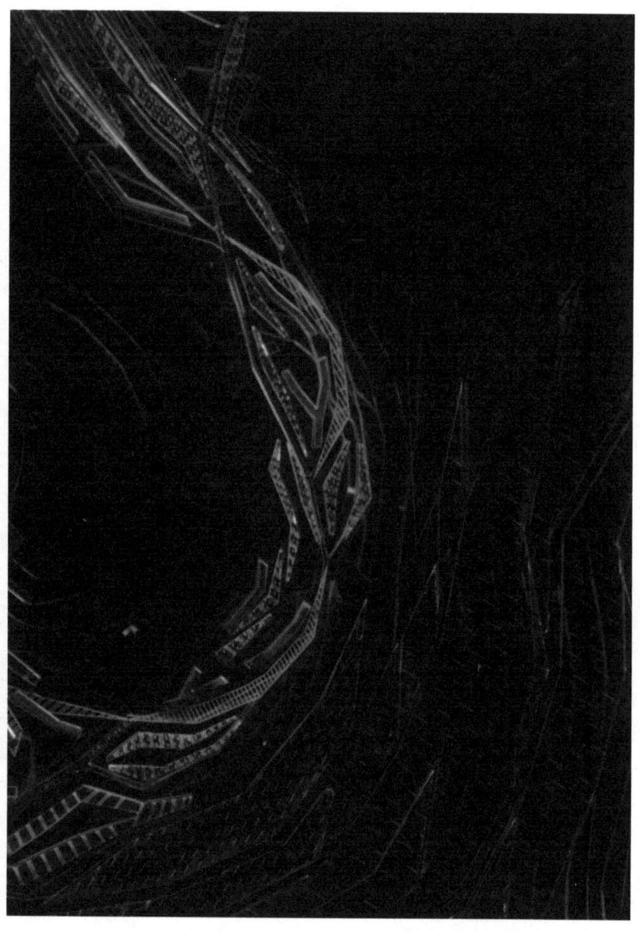

Prototype for Housing

Housing & Infrastructure

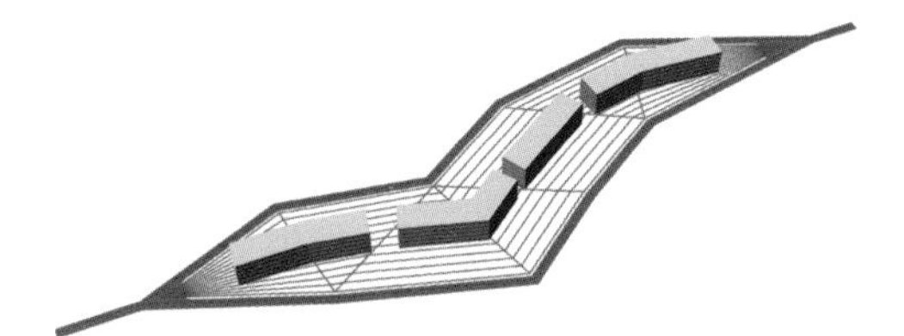

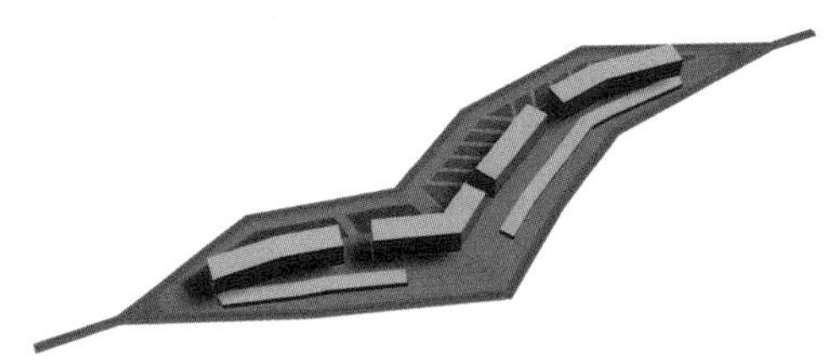

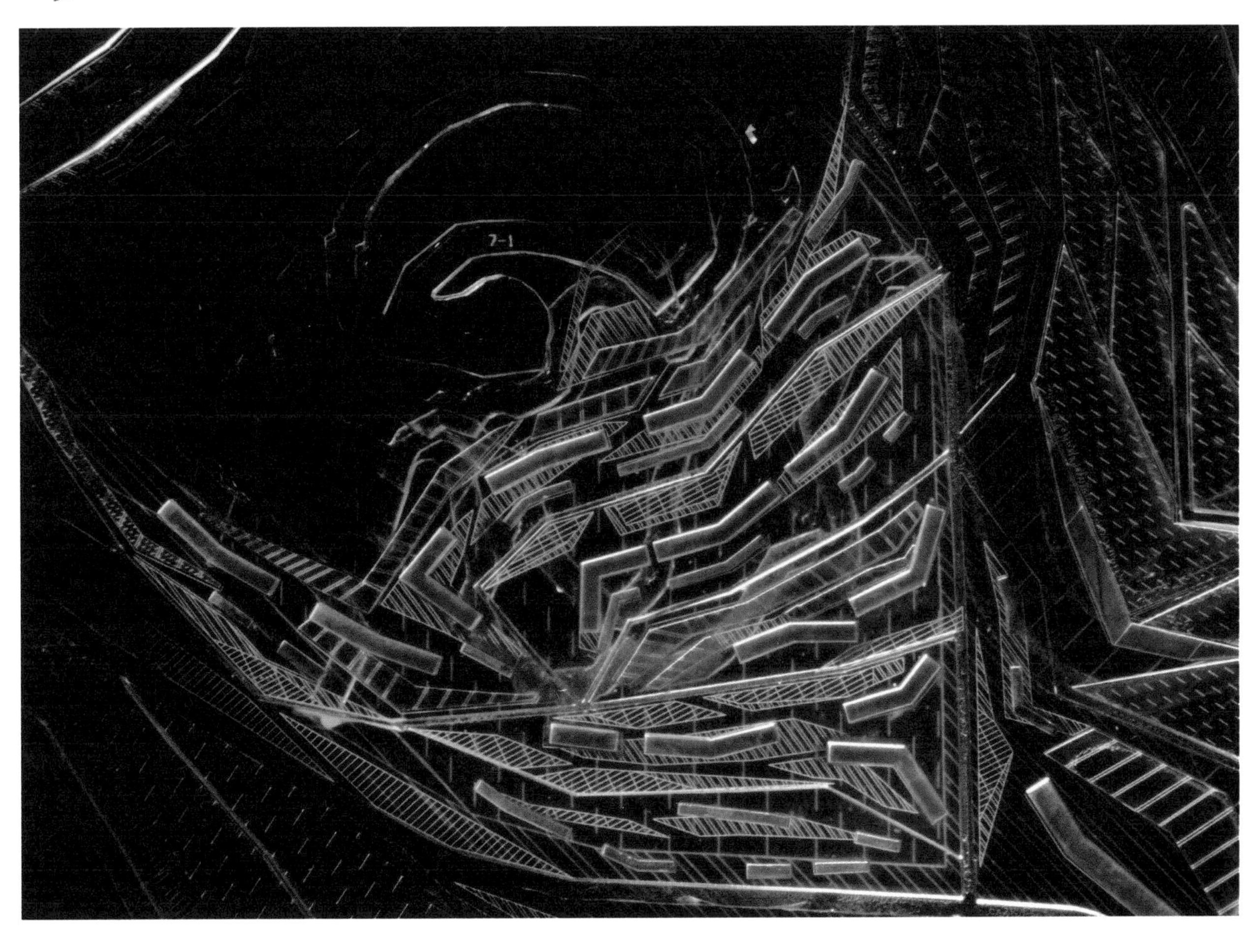

组 2

Team 2

姚亚男 盖若枚
Yao Yanan/Gai Ruomei

矿坑绿洲

Oasis in the Pit

湖北黄石矿坑是历史遗留的尺度巨大的采矿遗迹，是人工作用于自然所产生的独特的空间。伴随着开采的停止，矿坑也成为鲜有人问津的地方。而此次对其景观化的设计，希望能够使矿坑再一次成为人工与自然对话的空间。
本设计以场地的汇水为主要脉络，通过截水沟的利用将有限的雨水收集起来用于主要区域的灌溉，又最终汇集形成主要水面。通过人工手段对矿坑进行局部植被修复，同时也保留原有的场地记忆，形成新的人工与自然的共生。

Huangshi pit in Hubei is a large-scaled mining relics left over from history. It is also a unique space produced by power of both human and nature. As the mining operation is stopped, the pit is going to be a negative place. This studio work, we want to use approaches of landscape design to revive the pit, transforming it into a place where combined artificial and natural power.
Based on the rainwater catchment system, we design intercepting ditches to collect the limited rainwater for the irrigation of the main landscape areas. And eventually the water come together to form the wide water surfaces. We also restore the vegetation in the part of the pit, and retain the original memory of the pit at the same time. We want to form a new artificial and natural symbiotic situation in the pit.

姚亚男
Yao Yanan

盖若枚
Gai Ruomei

01 场地现状 Site Analysis

Plant water demand:

Dry field 4500m³/year · hm²

Paddy field 12000m³/year · hm²

Vegetable field 3000-7500m³/year · hm²

TERRACED WIDTH REFERENCE

Slope	Percentage (%)	Ridge height (M)	Field Width (M)	Slope distance (M)	Earthworks (M³)	Stonework (M³)	Employment (DAY)
3°	5.24	1.00	18,8	19.1	83.5	16.3	38~69
		1.50	28,3	28,7	[illegible]	17.4	41~85
		2,00	37,8	38,2	166.5	18,8	55~102
5°	8,75	1.00	11.2	11.5	92,0	27.4	30~100
		1.50	16.7	17,2	132,0	29.4	44~117
		2.00	[illegible]	23.0	172.0	31,8	57~136
7°	12,28	1,50	11,7	12,3	125,0	42,0	41~146
		2,00	15.7	18.4	166,5	45,0	55~168
		2,50	19.6	20,5	208,0	48,8	69~191
10°	17,63	1,50	8,1	8,6	133.0	60,6	45~196
		2.00	10,8	11,5	175,0	65,7	58~222
		2,50	13.4	14,4	209.0	71,7	69~249
15°	26,79	2,00	6.9	7,7	180,0	103.0	60~318
		2,50	8,6	9.7	218,0	111,0	72~351
		3 00	10,2	11.6	250,0	122,0	83~388

Mesh Principle

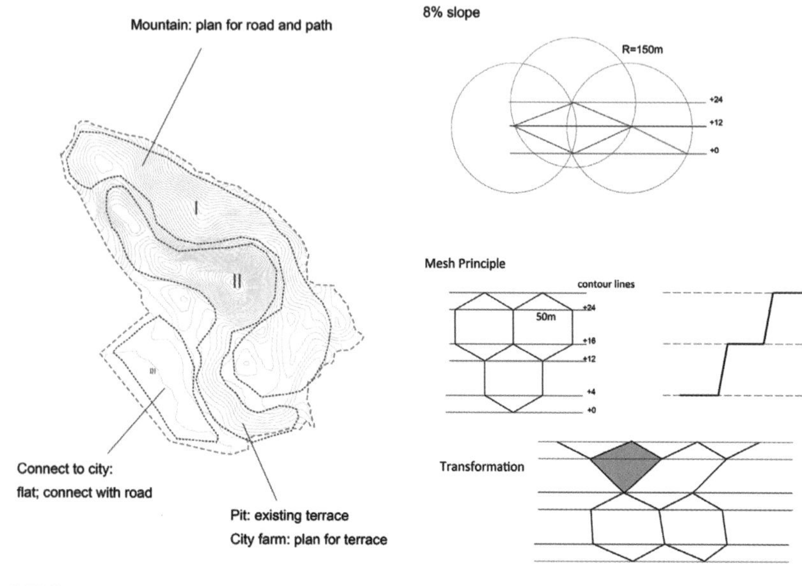

Basic Research

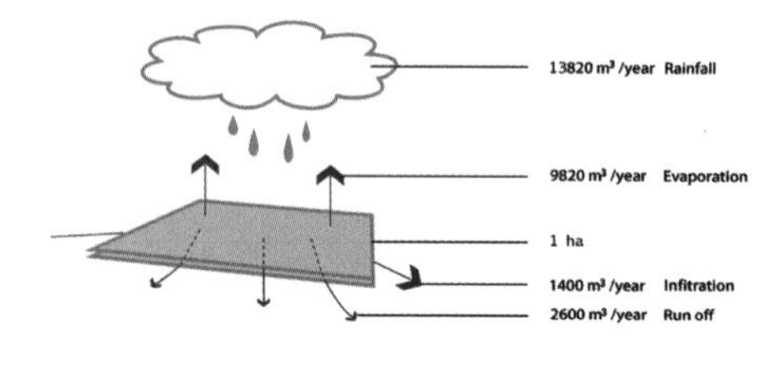

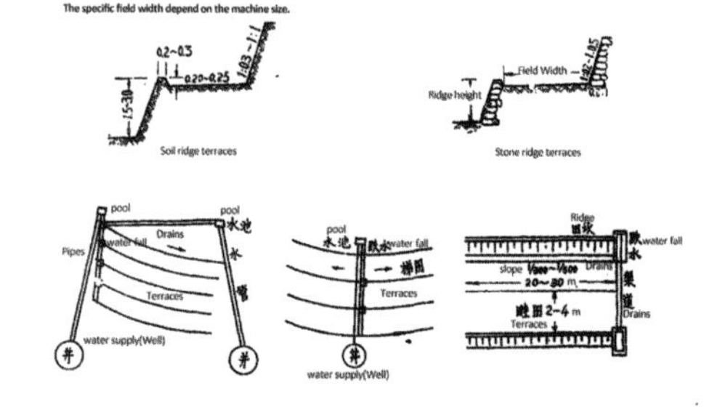

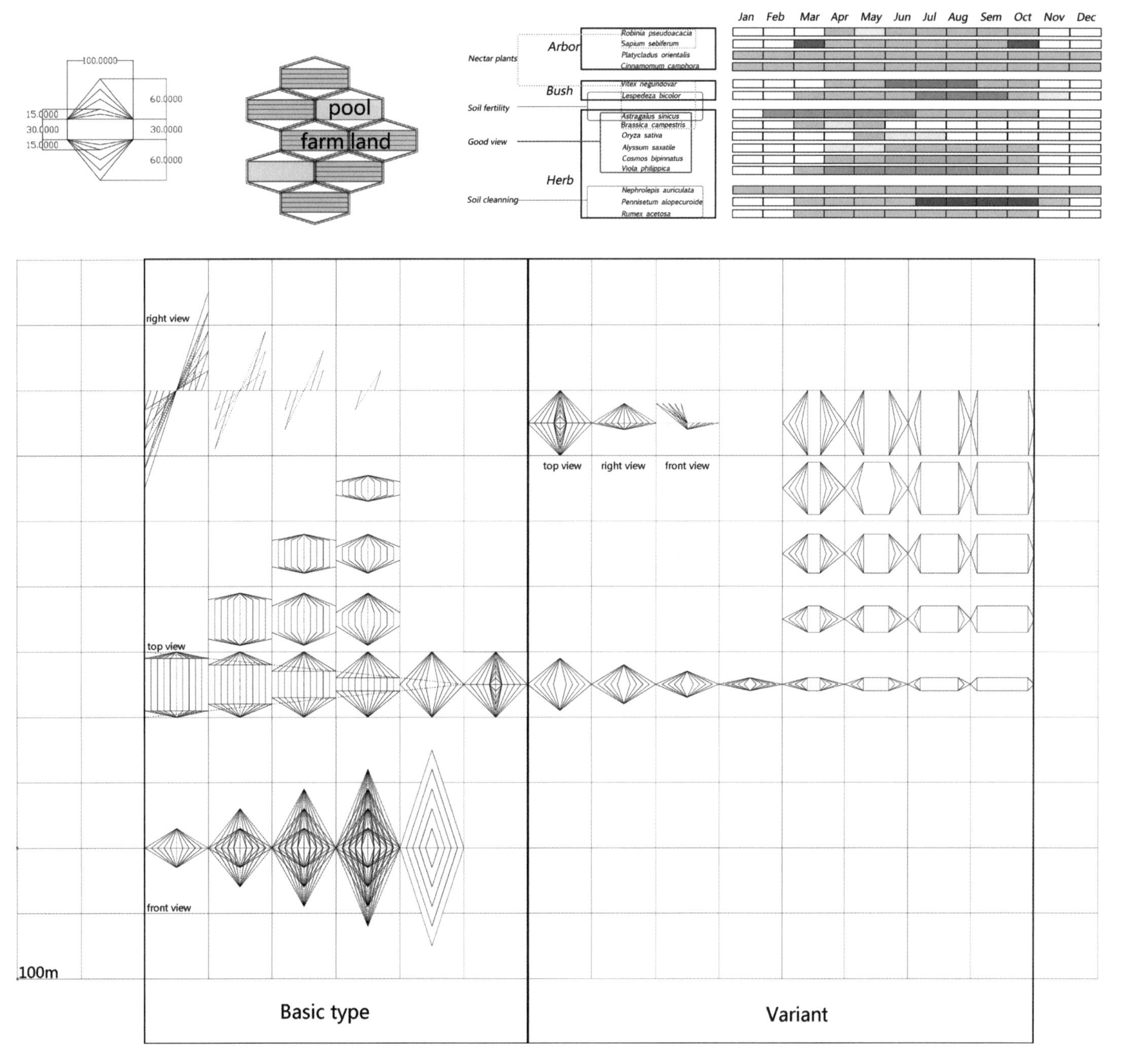
100.0000
60.0000
15.0000
30.0000
30.0000
15.0000
60.0000
pool
farm land
Jan Feb Mar Apr May Jun Jul Aug Sem Oct Nov Dec
Nectar plants
Arbor
Bush
Herb
Soil fertility
Good view
Soil cleanning
Robinia pseudoacacia
Sapium sebiferum
Platycladus orientalis
Cinnamomum camphora
Vitex negundovar
Lespedeza bicolor
Astragalus sinicus
Brassica campestris
Oryza sativa
Alyssum saxatile
Cosmos bipinnatus
Viola philippica
Nephrolepis auriculata
Pennisetum alopecuroide
Rumex acetosa
right view
top view
front view
top view
right view
front view
100m
Basic type
Variant

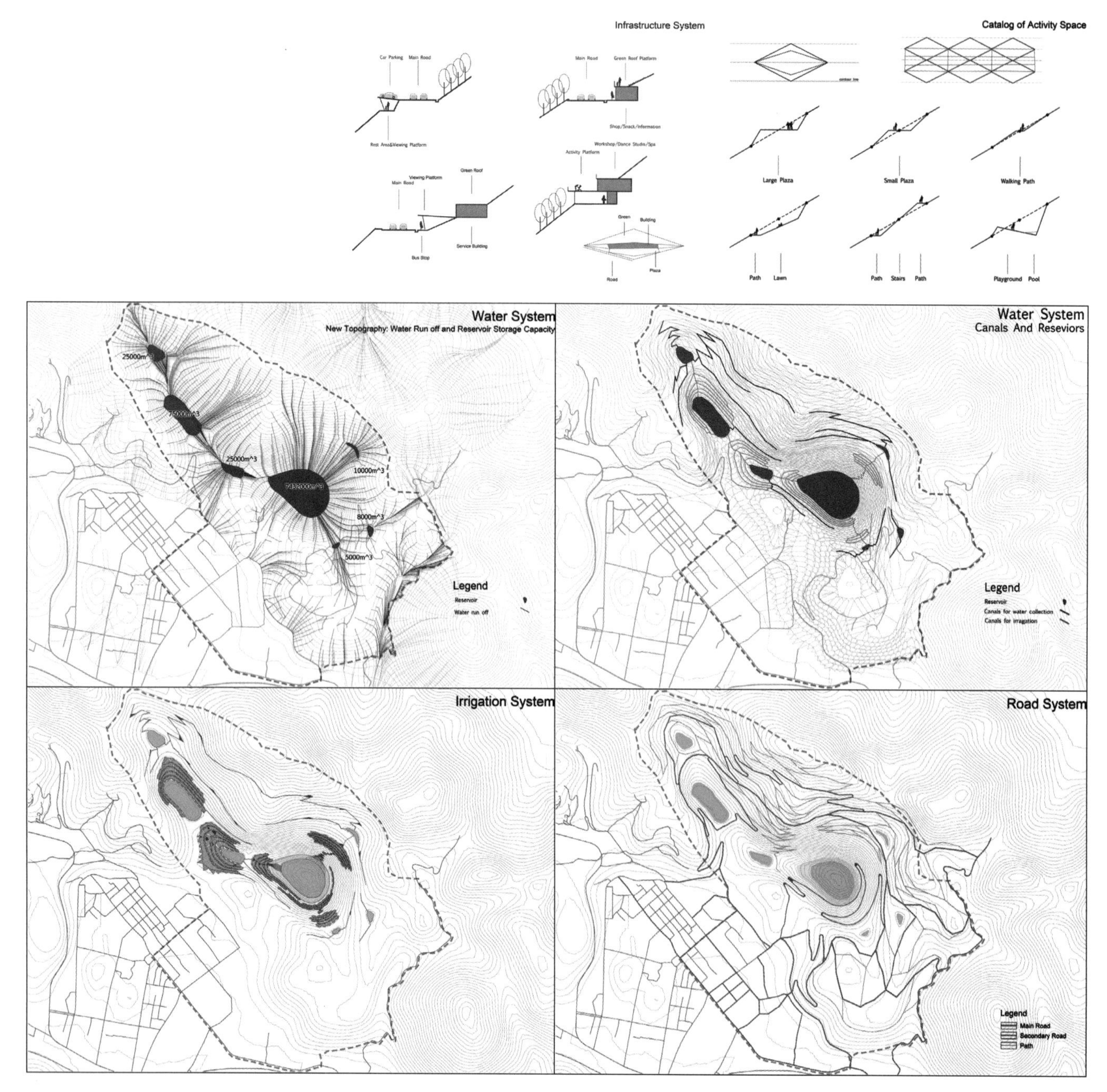
Infrastructure System
Catalog of Activity Space
Car Parking
Main Road
Rest Area&Viewing Platform
Green Roof Platform
Shop/Snack/Information
Workshop/Dance Studio/Spa
Activity Platform
Viewing Platform
Green Roof
Service Building
Bus Stop
Green
Building
Road
Plaza
Large Plaza
Small Plaza
Walking Path
Path
Lawn
Stairs
Playground
Pool
Water System
New Topography: Water Run off and Reservoir Storage Capacity
25000m^3
10000m^3
8000m^3
5000m^3
Legend
Reservoir
Water run off
Water System
Canals And Reseviors
Canals for water collection
Canals for irrigation
Irrigation System
Road System
Main Road
Secondary Road

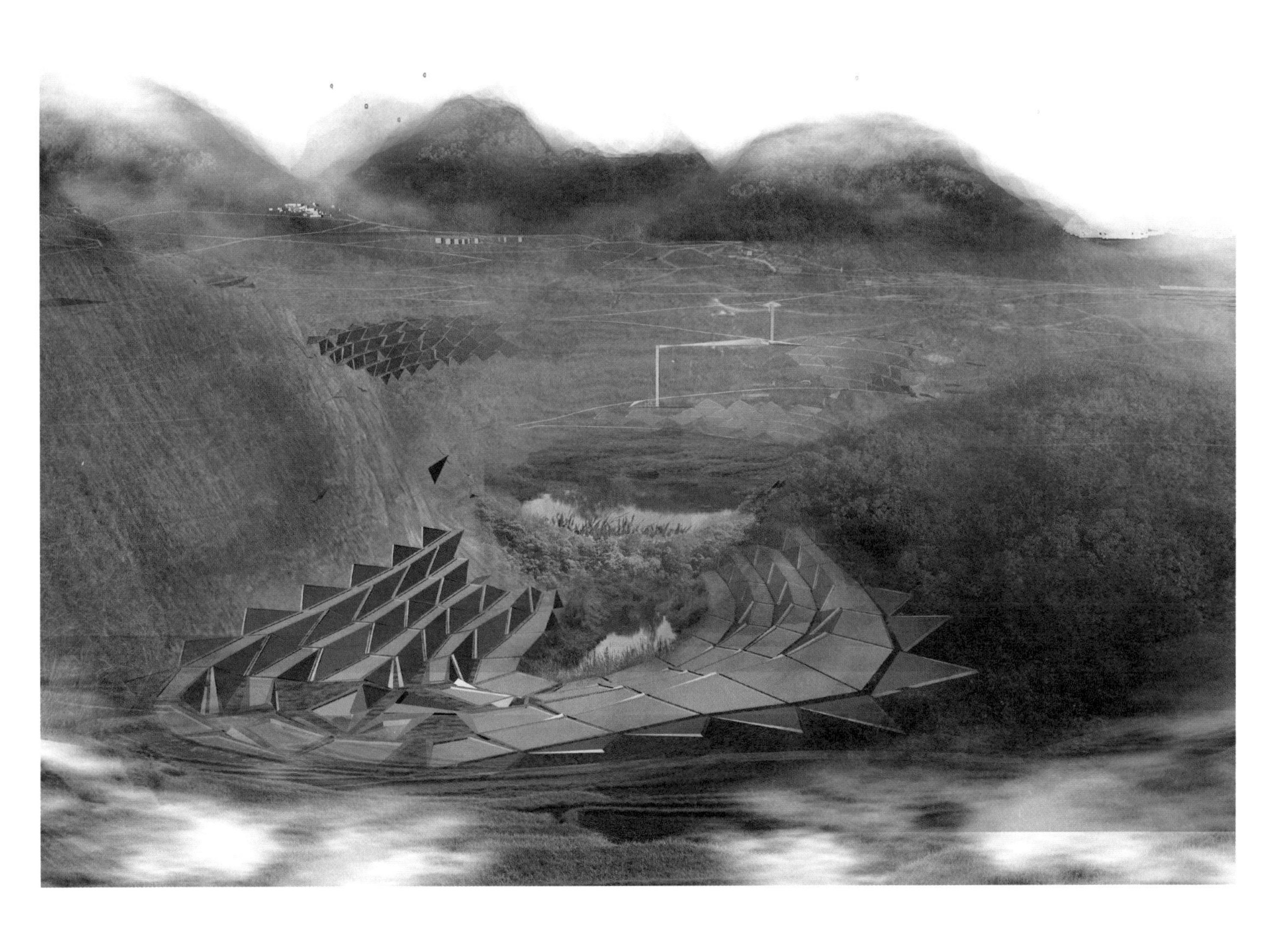

04 模型展示 Design Model

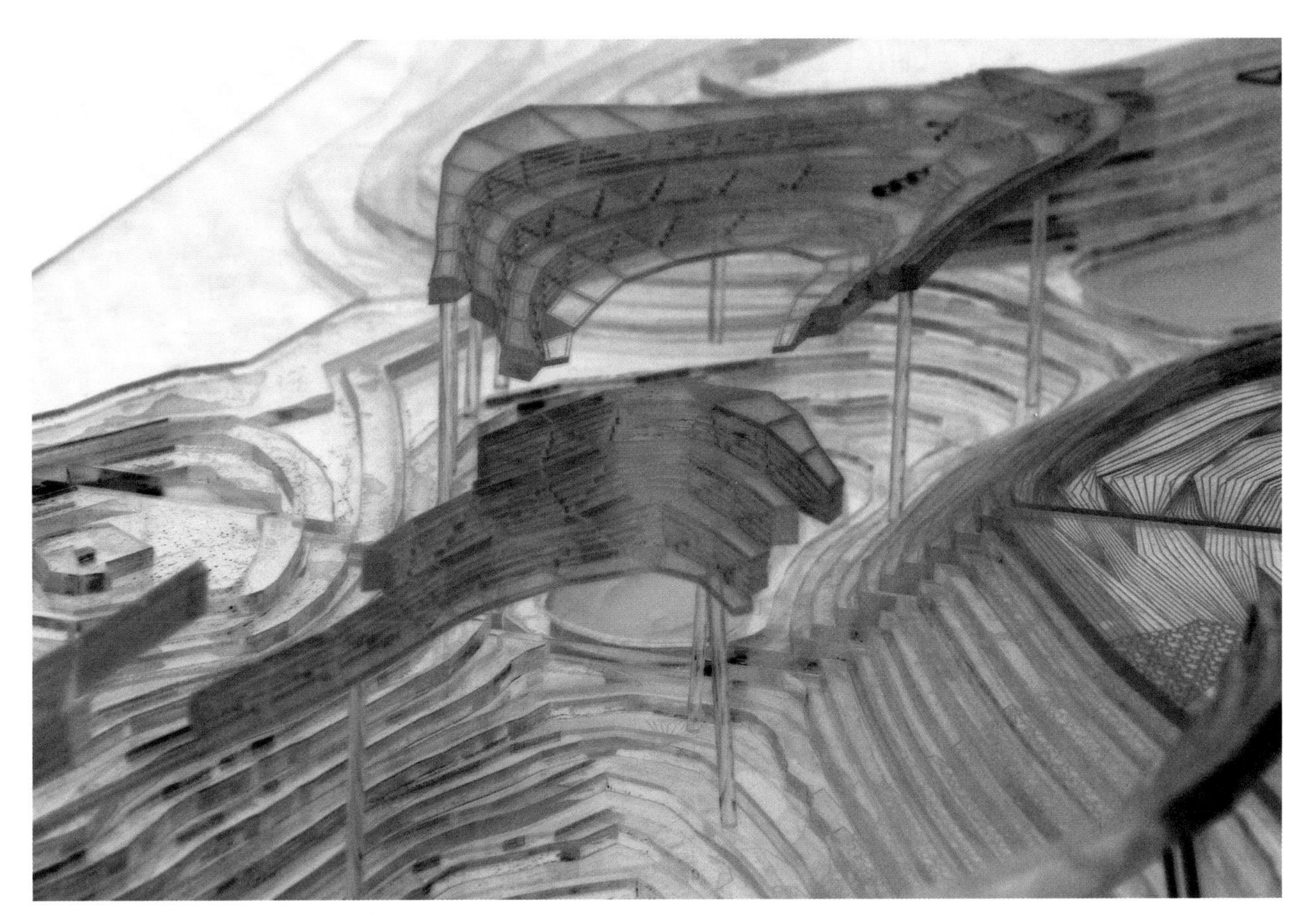

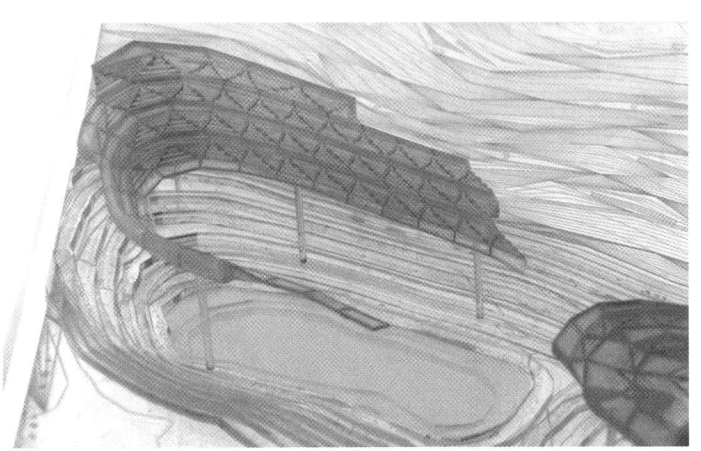

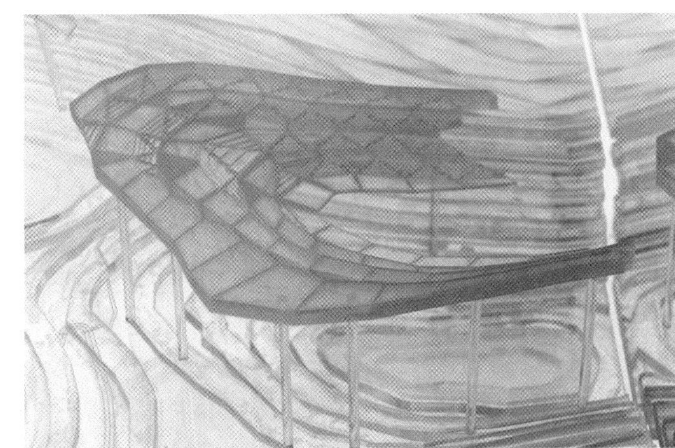

组 3
Team 3

李芸芸　张倩玉　张益章
Li Yunyun/Zhang Qianyu/Zhang Yizhang

方案定位为中药种植、生产和物流产业。根据场地对外交通条件，将生产节点和流程置入空间，并由地质条件确定种植类型和区域。对产业进一步研究后，得出种植、晾晒、生产、运输等功能在空间上的关系。将场地根据路网和地形进行网格化（meshing）处理，再由产量和运输距离等因素的量化分析一步步推出各类加工建筑面积。生态上通过对降水的分析，将水净化和灌溉系统纳入到网格的系统中。最后归纳出几种原型应用在不同类型的网格中。

The project set target in the planting, production and logistics industry of Chinese medicine. According to the site external traffic conditions, the production node and process is set into space, and the type of planting and regional is determined by the geological conditions . Through deeper researching on the production process, the functions of planting, drying, transportation are located in space. The site is meshed according to the road network and terrain, and then various types of processing buildings area are calculated through quantitative analysis of the production and transportation distance. Through the analysis of precipitation, water purification and irrigation systems are incorporated into the meshing system. Finally, several prototype applications are summarized in different types of mesh.

李芸芸
Li Yunyun

张倩玉
Zhang Qianyu

张益章
Zhang Yizhang

01 场地现状 Site Analysis

Area: 300 hectares

Population:

cultivation:300 households

processing & logistics: 4500 households

Production value:

6+0.5 (cultivation)hundred billion, 45000/capita (GDP)

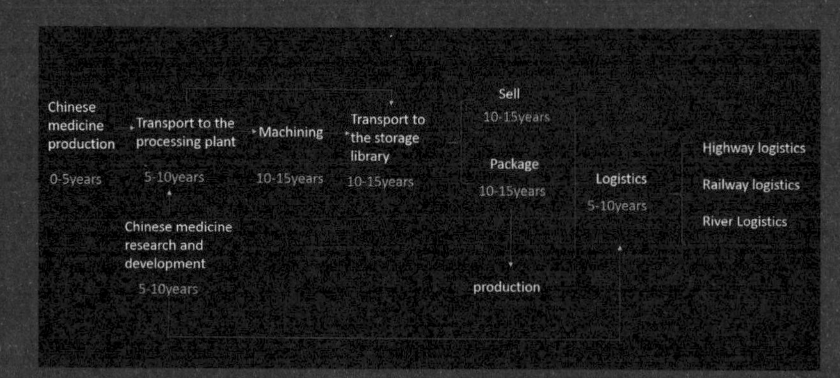

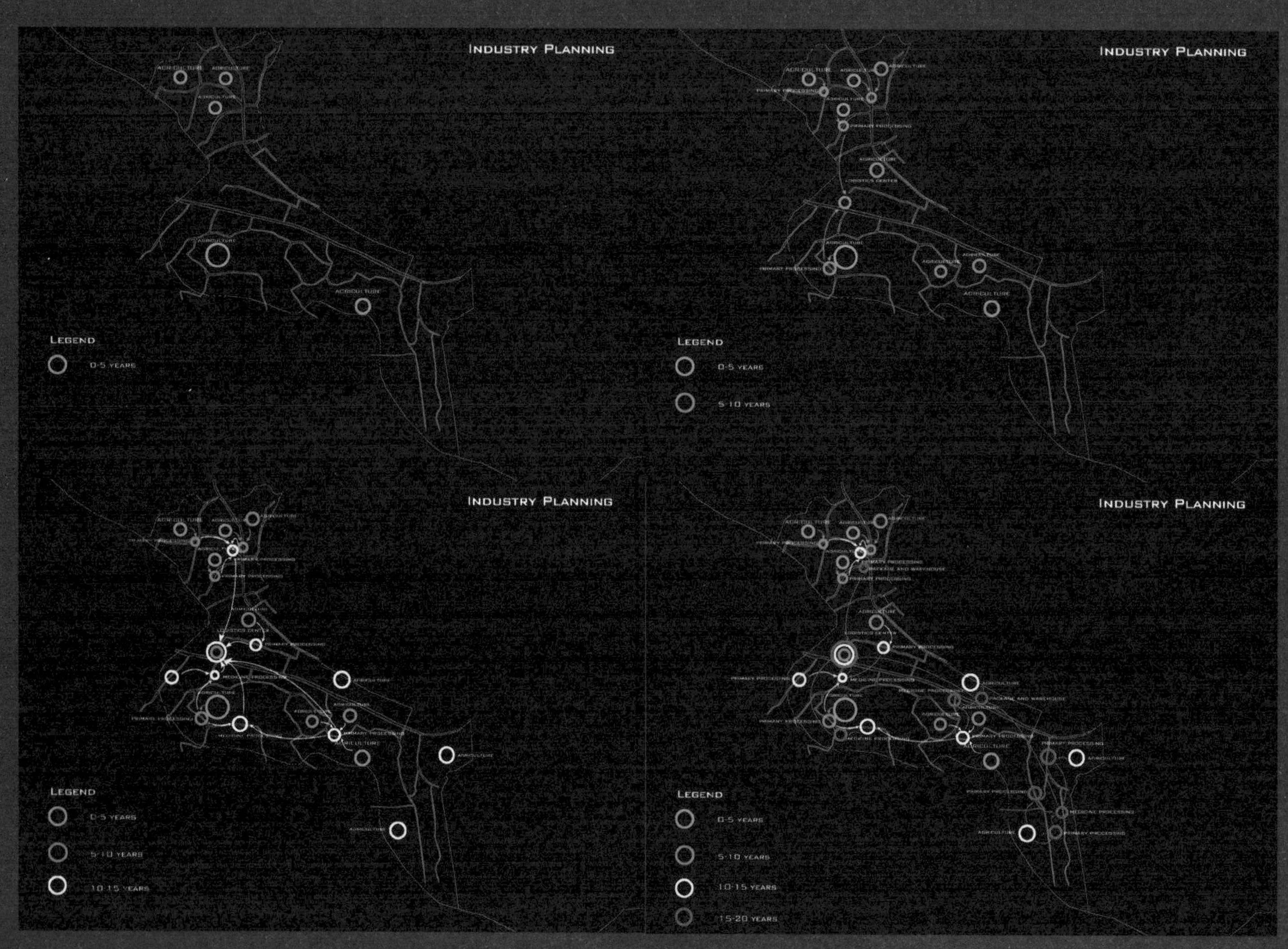

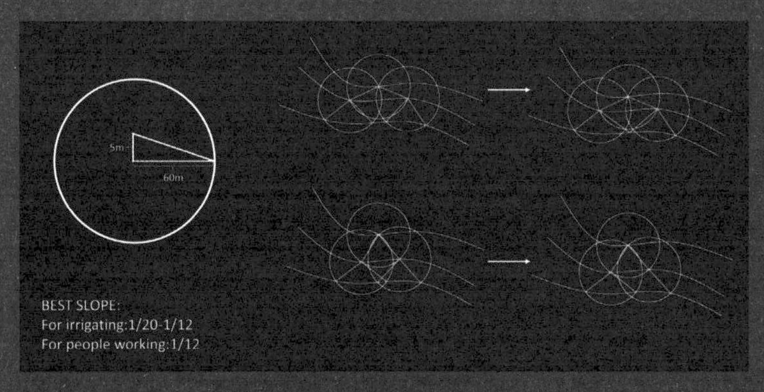
5m
60m
BEST SLOPE:
For irrigating:1/20-1/12
For people working:1/12

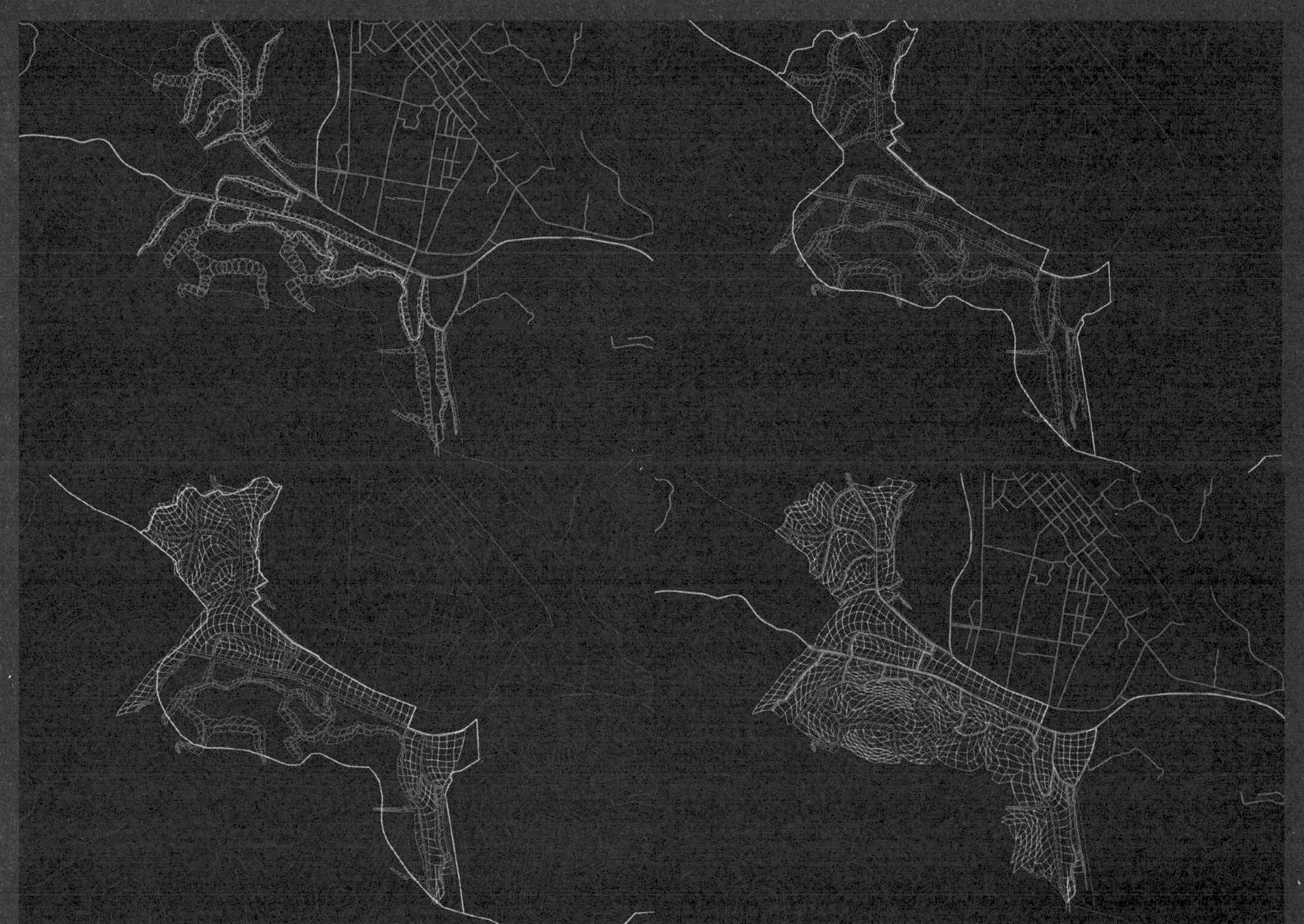

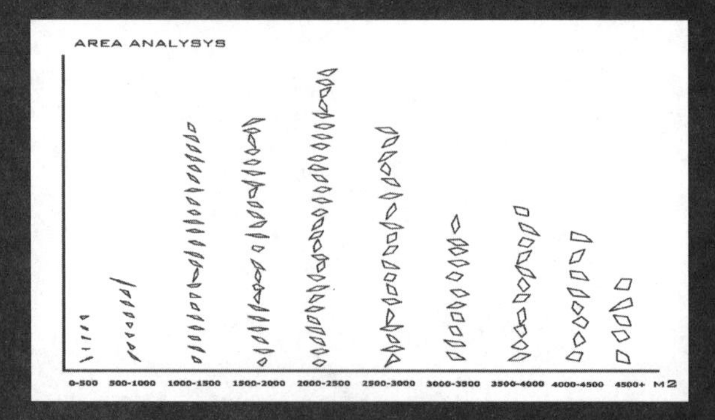
AREA ANALYSYS
0-500
500-1000
1000-1500
1500-2000
2000-2500
2500-3000
3000-3500
3500-4000
4000-4500
4500+
M2

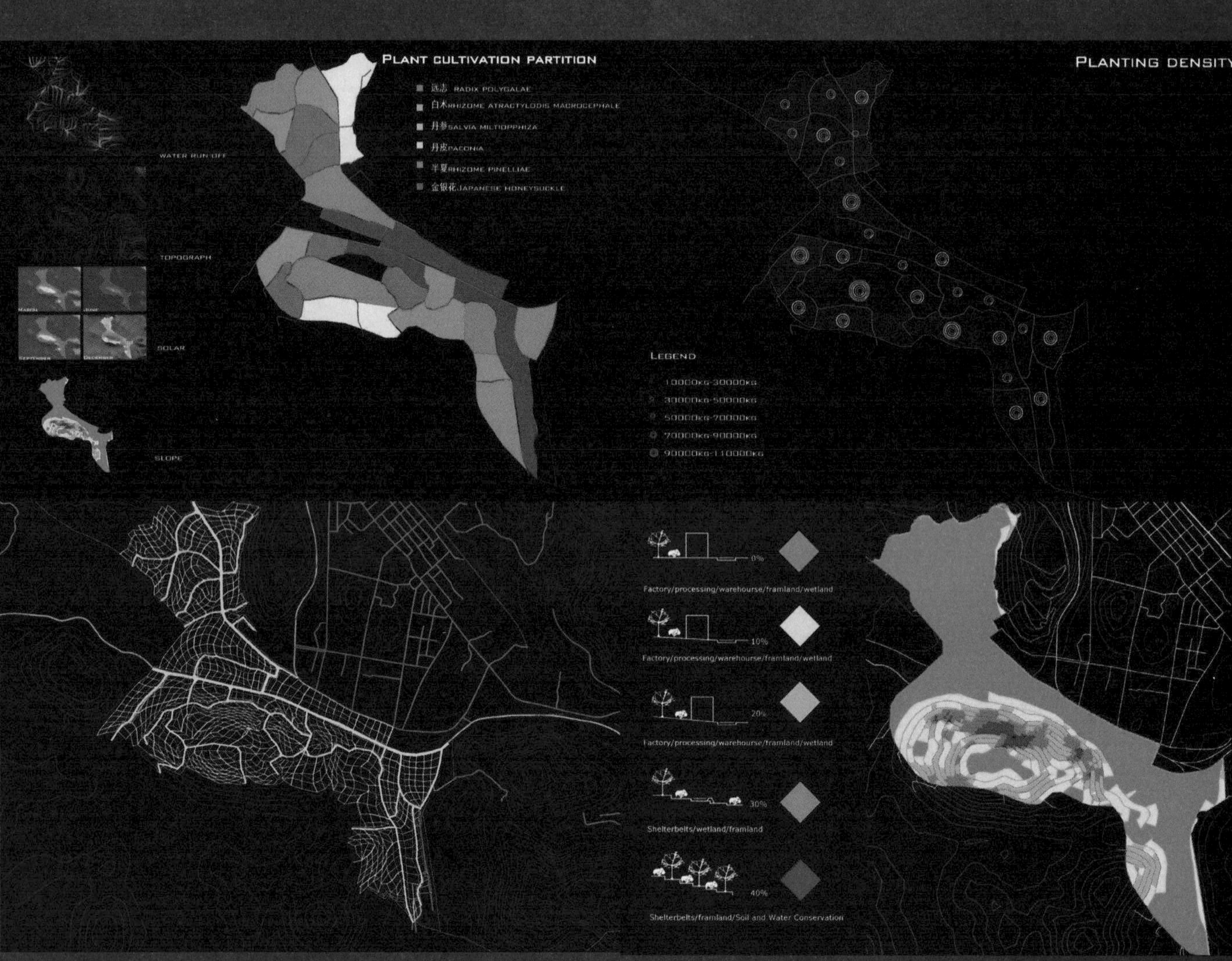
PLANT CULTIVATION PARTITION
远志 RADIX POLYGALAE
白术RHIZOME ATRACTYLODIS MACROCEPHALE
丹参SALVIA MILTIOPPHIZA
丹皮PACONIA
半夏RHIZOME PINELLIAE
金银花JAPANESE HONEYSUCKLE
WATER RUN-OFF
TOPOGRAPH
SOLAR
SLOPE
PLANTING DENSITY
LEGEND
10000KG-30000KG
30000KG-50000KG
50000KG-70000KG
70000KG-90000KG
90000KG-110000KG
0%
Factory/processing/warehourse/framland/wetland
10%
Factory/processing/warehourse/framland/wetland
20%
Factory/processing/warehourse/framland/wetland
30%
Shelterbelts/wetland/framland
40%
Shelterbelts/framland/Soil and Water Conservation

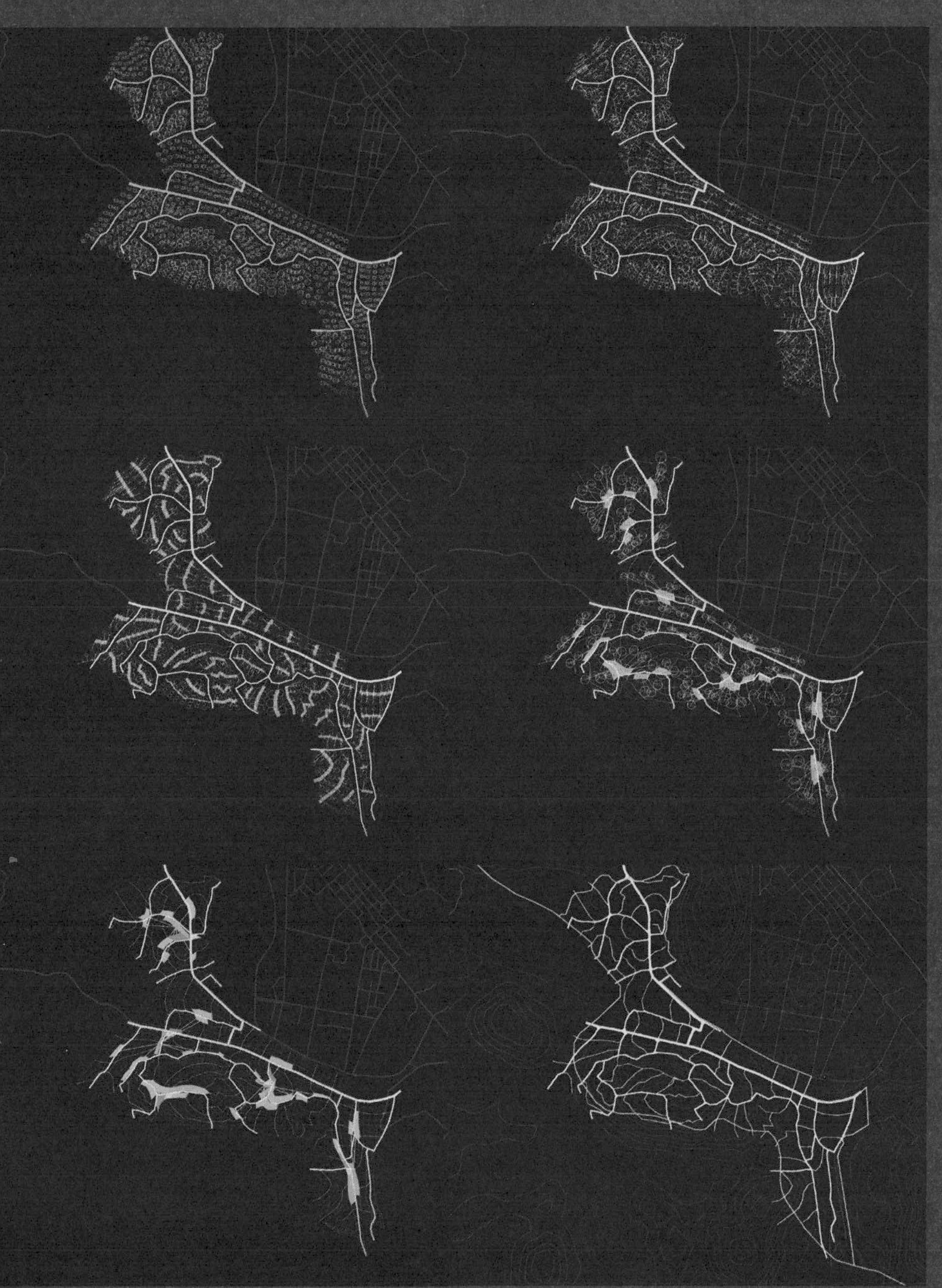

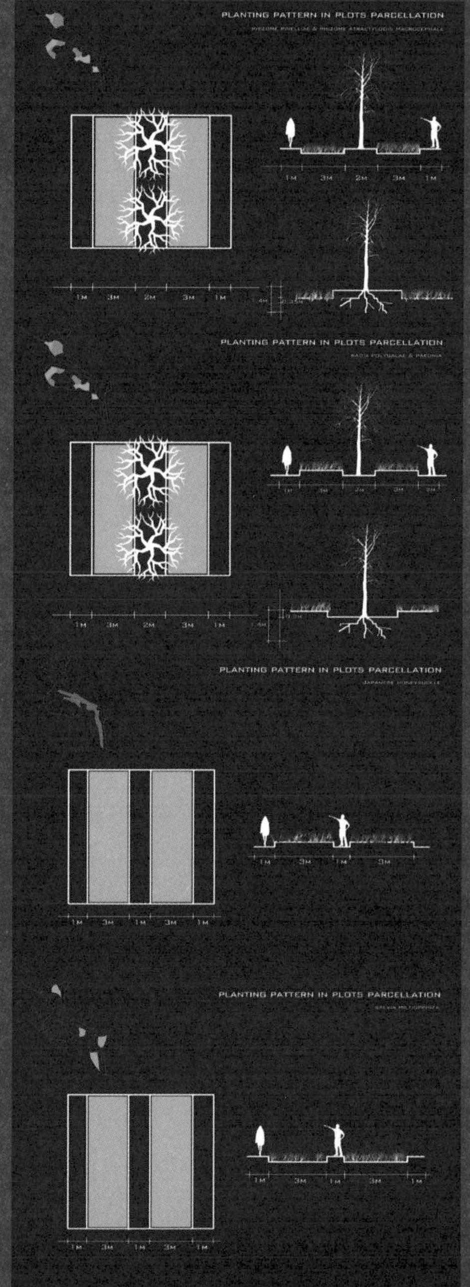

PLANTING PATTERN IN PLOTS PARCELLATION
1M 3M 2M 3M 1M
1M 3M 2M 3M 1M
PLANTING PATTERN IN PLOTS PARCELLATION
1M 3M 2M 3M 1M
PLANTING PATTERN IN PLOTS PARCELLATION
1M 3M 1M 3M
1M 3M 1M 3M 1M
PLANTING PATTERN IN PLOTS PARCELLATION
1M 3M 1M 3M 1M
1M 3M 1M 3M 1M

02 总体设计 General Design

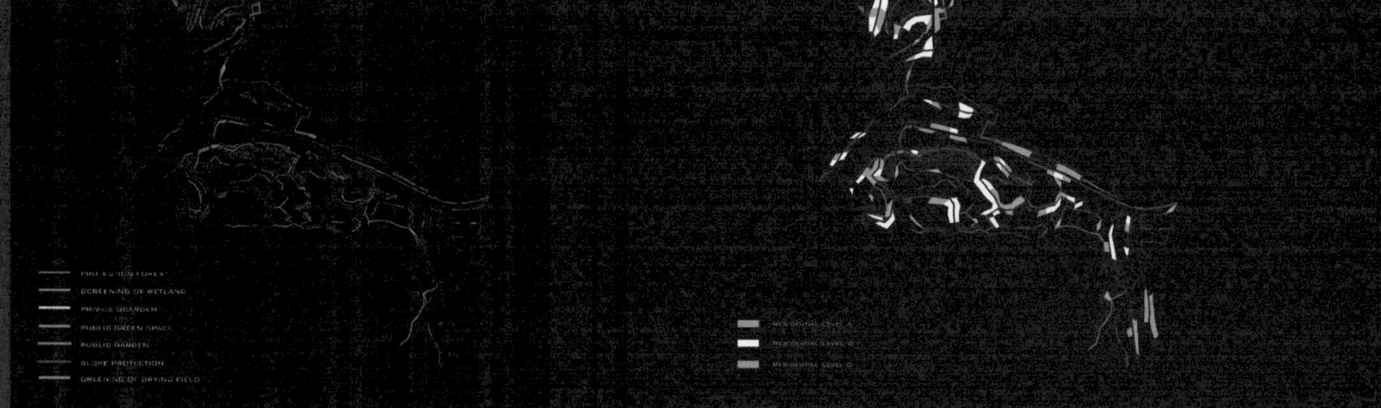

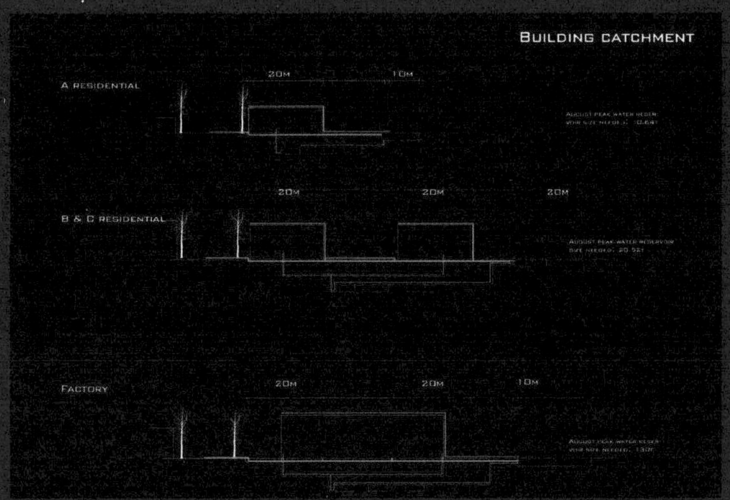
BUILDING CATCHMENT
A RESIDENTIAL
20M
10M
B & C RESIDENTIAL
20M
20M
20M
FACTORY
20M
20M
10M

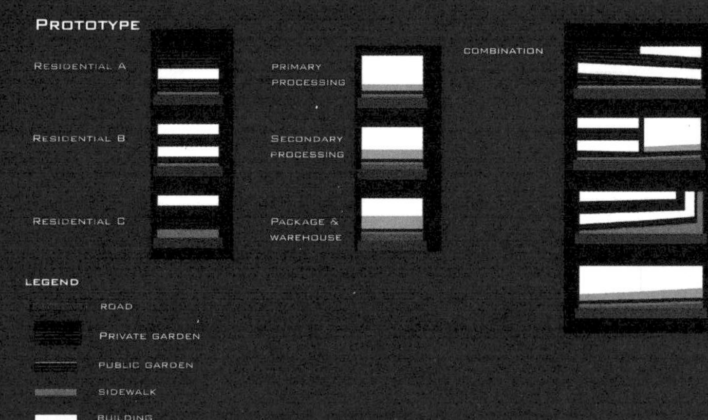
PROTOTYPE
RESIDENTIAL A
RESIDENTIAL B
RESIDENTIAL C
PRIMARY PROCESSING
SECONDARY PROCESSING
PACKAGE & WAREHOUSE
COMBINATION
LEGEND
ROAD
PRIVATE GARDEN
PUBLIC GARDEN
SIDEWALK
BUILDING

WATER COLLECT IN RESIDENTIAL AREA
47083M3
33412M3
50119M3
37969M3
INHABITANT: 59PEOPLE EACH TILE
AREA: 1800M2
BUILDING: 1UNIT
WATER COLLECT AREA:225(ROOF)+900 (GARDEN)= 1125M2
AVERAGE WATER COLLECT AMOUNT: 1518.75M3 / YEAR
AVERAGE GRAY WATER WASTE: 2595 M3/YEAR
WATER COLLECT IN RESIDENTIAL AREA
38880M3
24300M3
19440M3
4840M3
INHABITANT: 108PEOPLE EACH TILE
AREA: 3600M2
BUILDING: 2UNIT
WATER COLLECT AREA:600(ROOF)+1200 (GARDEN)=1800M2
AVERAGE WATER COLLECT AMOUNT: 2430M3 / YEAR
AVERAGE GRAY WATER WASTE: 4752M3/YEAR

Buildings
Greening system
Road system
Farmland

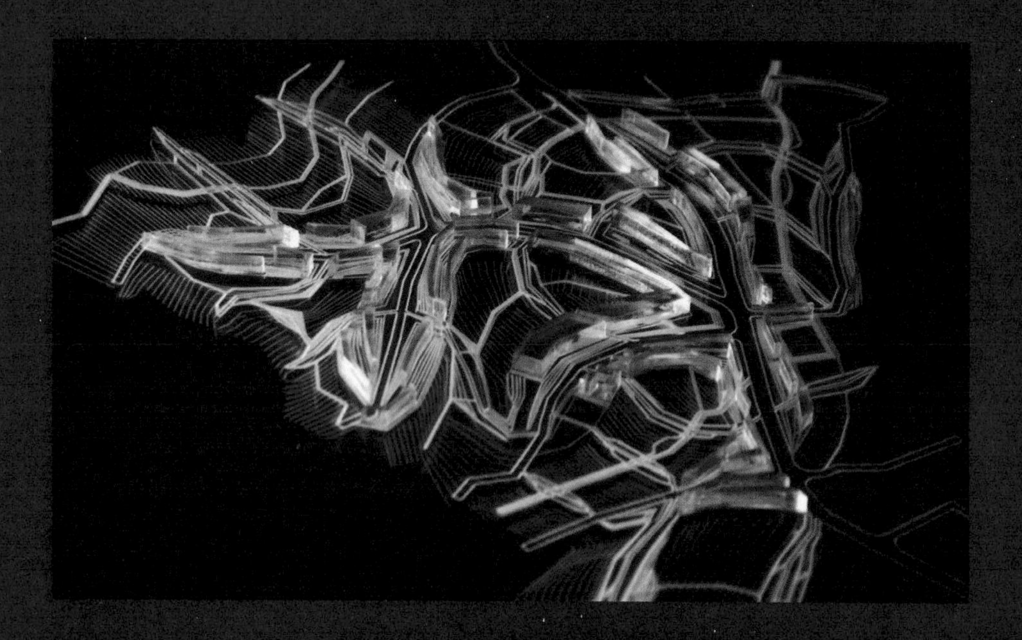

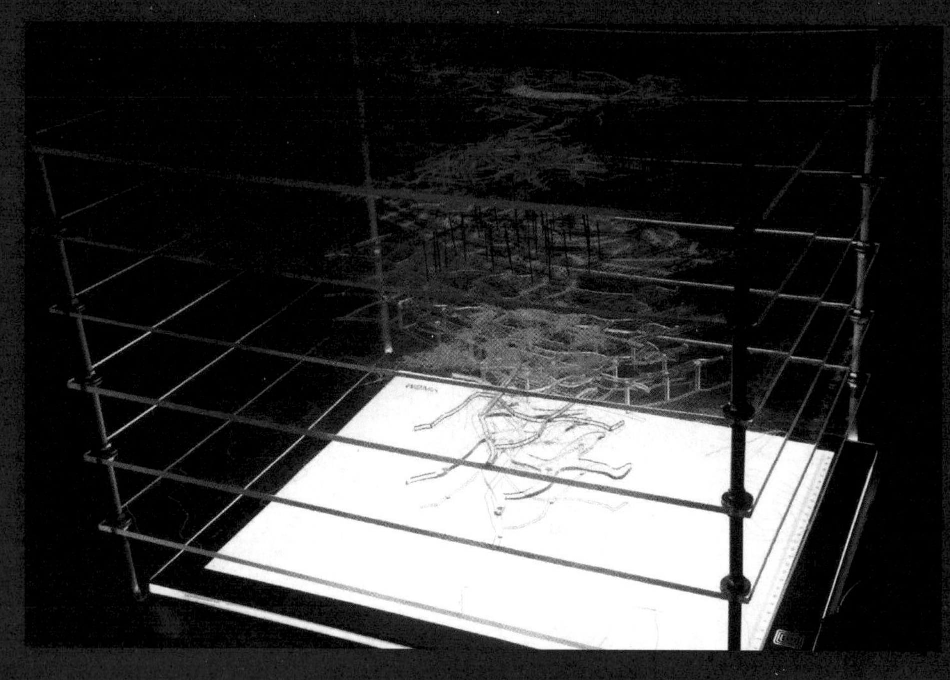

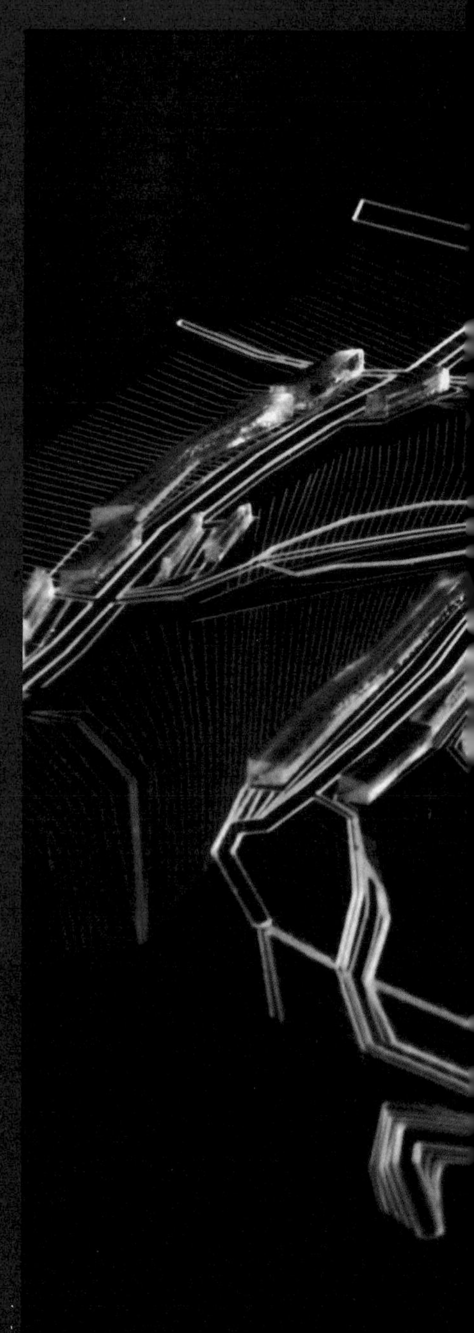

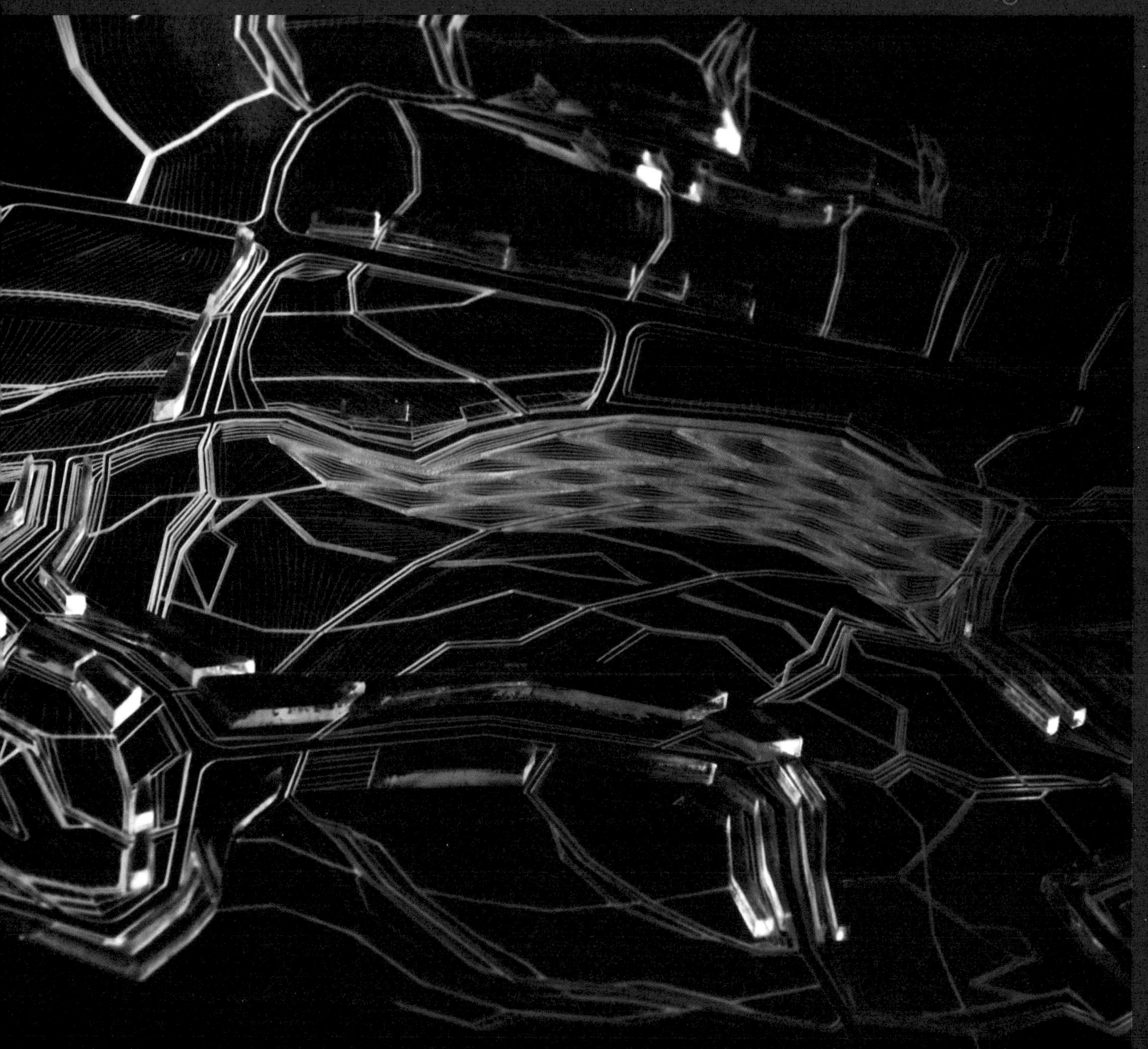

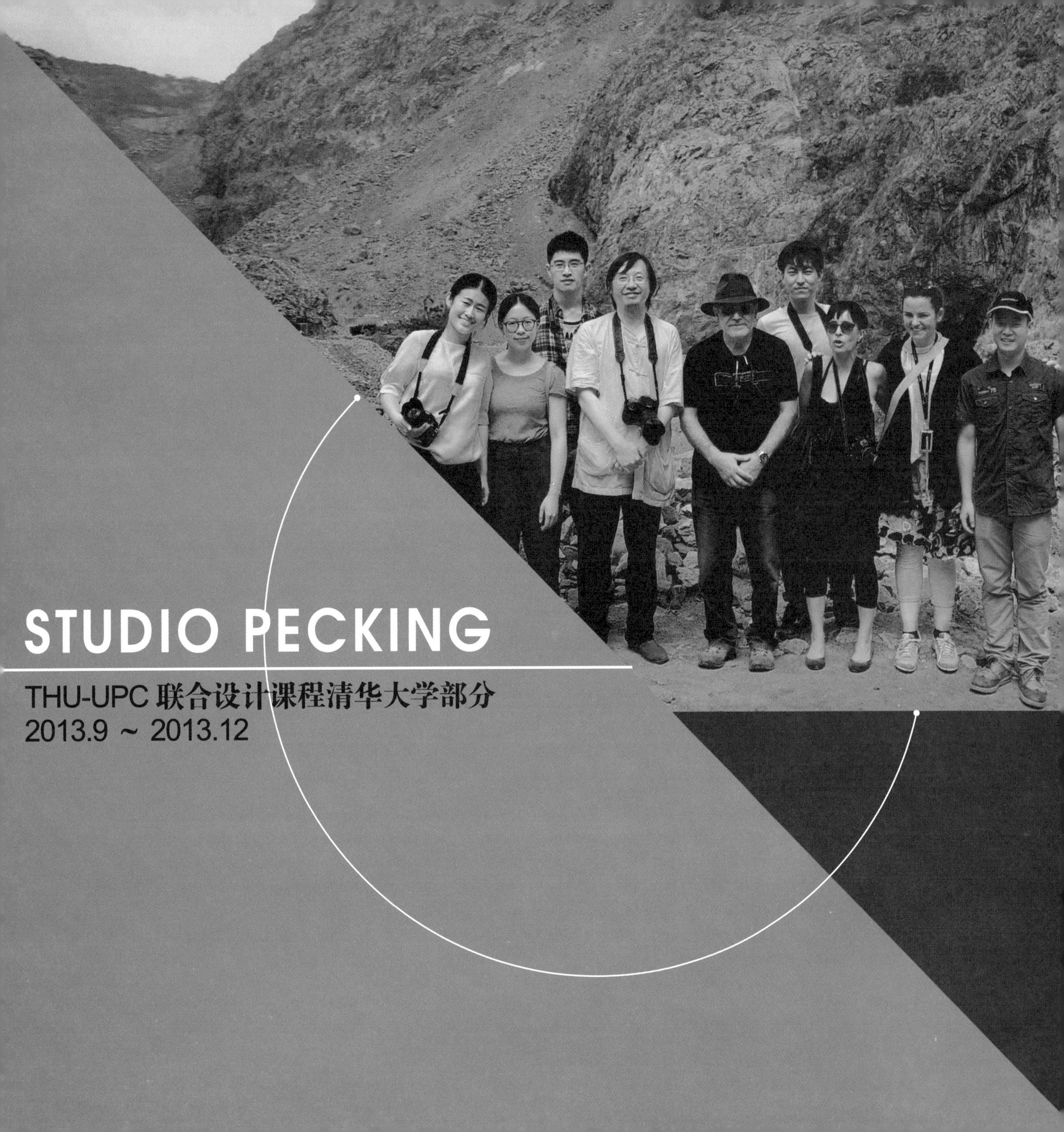

STUDIO PECKING

THU-UPC 联合设计课程清华大学部分
2013.9 ~ 2013.12

时间的痕迹

Trace of Time

陈美霞　李凯　赵佳萌　杨帆

Chen Meixia/Li Kai/Zhao Jiameng/Yang Fan

巨大的天坑是黄石市工业遗址中最重要的场地之一。该场地面临资源枯竭、经济结构改变。转型旅游是这里发展的一个机遇。保留工业历史遗迹是开发旅游的立足点。对于这座城市来说，这是未来的财富。

此设计所做的事情就是通过触摸时间的轨迹，去寻找这座被废弃矿山的潜在可能性。在过去，天坑底部代表了矿产资源枯竭。但未来，矿坑将作为城市旅游资源，并且在紧急情况下将作为城市备用水库为城市服务。本设计的旅游规划定位与水系规划设计均隐喻了铁山区新的希望。

Its huge sinkhole is one of the most important Huangshi industrial remains. The site faces the resource depletion the economic structural change.Transformation for tourism is an opportunity of the city.

What the design do is to find out the potential possibility of the abandoned mountain mine trough touching the trace of time.In the past,the bottom of the pit represented for the exhausted mining. In the future ,the pit will become a tourist resource, as well as a city backup reservoir for emergency situations. The tourism planning positioning and the water system planning are a metaphor for the new hope of this Iron Mountain.

01 场地现状 Site Analysis

场地简介 Brief Introduction of the Site

我们进入矿坑中心，希望找到一些能体现场地环境的特质。从场地边缘至中心矿坑，有从自然到人工的痕迹演变；从主入口到中心矿坑，环境由富有历史展示的区域逐渐变为空旷的废弃的矿坑；基地中同是两个巨大下陷的区域，一个是东方山水库，一个是矿坑本体，这两者产生强烈的对比。

We enter the space center the empty pit, hope to find some identity to illustrate the characteristics of context .From the periphery and gradually into the pit, we found the venue gradually shift manually from natural ; From the main entrance of the city gradually to the pit, the venue changes from a rich historical landscape space to an empty abandoned pit space ; similar analogy in space, we found there has been strong contrast between the an empty pit with the full neighboring East Hill Reservoir.

区位分析 Location

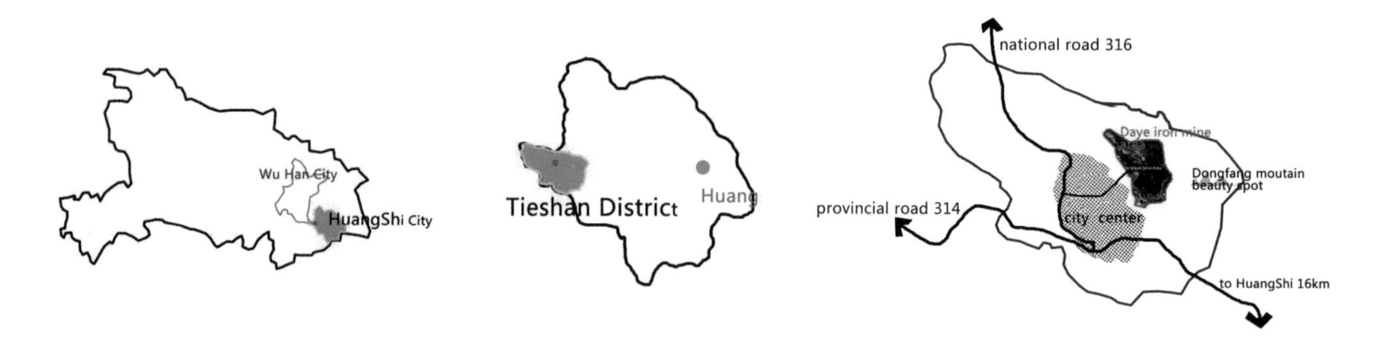

场地特质演变 SITE IDENTITY EVOLUTION

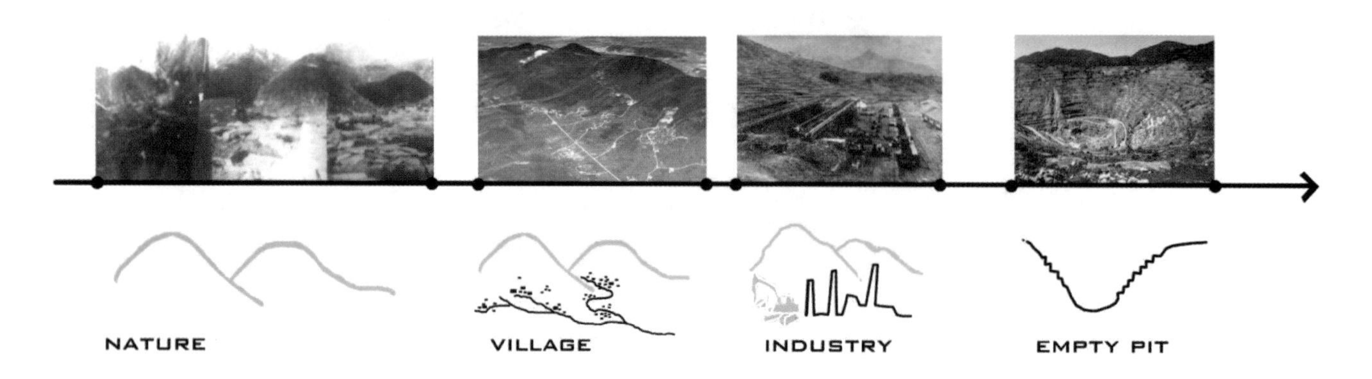

时间箭下坑路线概念 Concept Of Trailway Down To The Pit

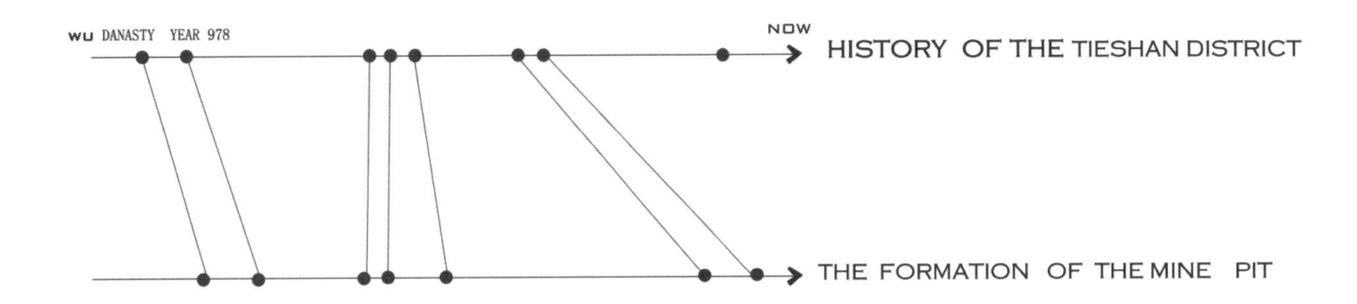

历史沿革 History

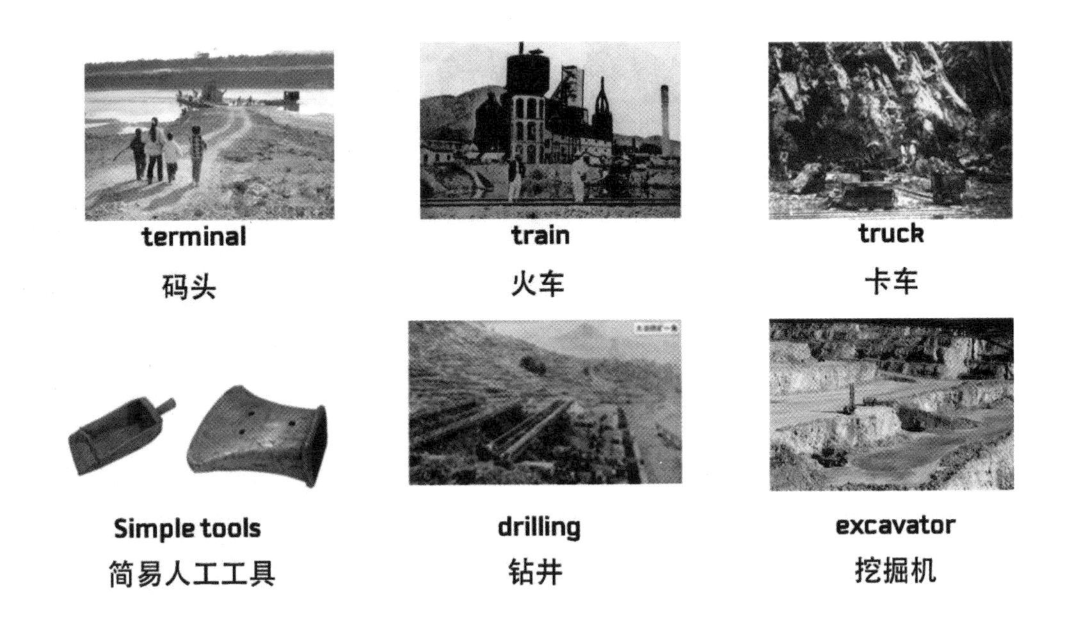

01 场地现状 Site Analysis

基地分析与感知 SITE ANALYSIS AND PERCEPTION

时间和空间在超常尺度规模的环境中扭曲，变形，重叠。在人类可以感知的范围内，我们可以得到的印象仅仅是一个整体的印象，这样的印象是静态的。最后，我们能够决定的只是如何观察。

Time and space in the scale of the environment is twisted, deformation. Within the scope of the human can sense, we can get the impression that is just a whole picture of impression, the impression is static. Finally, we can decide, only how to pass and how to watch.

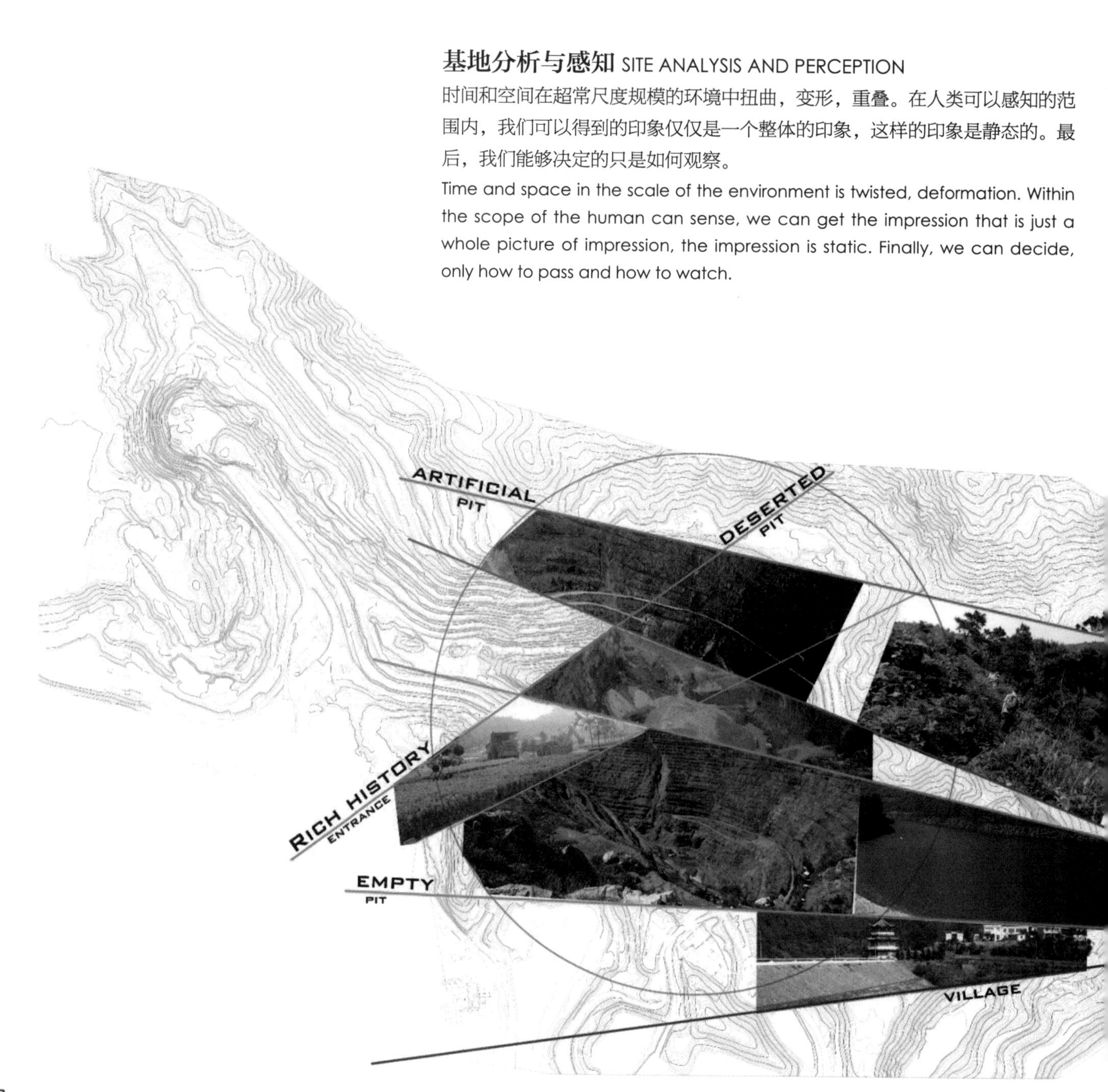

02 理念构思 Concept

概念分析图

Concept Diagram

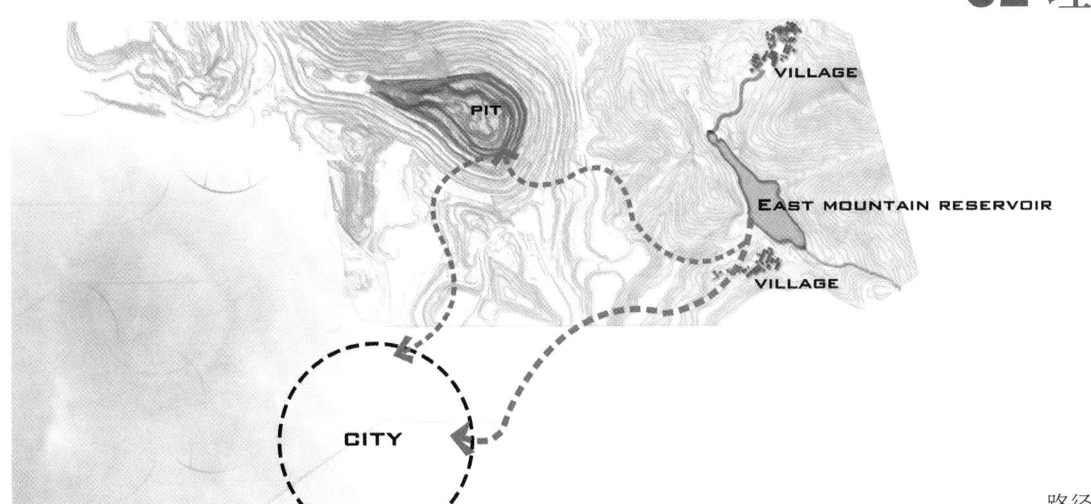

道路、水系统设置

Tour Line And Water System

路径的设计必须简洁，在矿坑环境中要简单明了。也因此路径在整个环境中异质性是明确的，作为一个独立的外部系统连接到矿坑本体。我们的设计重点是路径的不断转换和综合可利用的现状，最终目的是形成独立的设计语言。

We enter the space center the empty pit, hope to find some identity to illustrate the characteristics of context. From the periphery and gradually into the pit, we found the venue gradually shift manually from natural ;

From the main entrance of the city gradually to the pit, the venue changes from a rich historical landscape space to an empty abandoned pit space ; similar analogy in space, we found there has been strong contrast between the an empty pit with the full neighboring East Hill Reservoir.

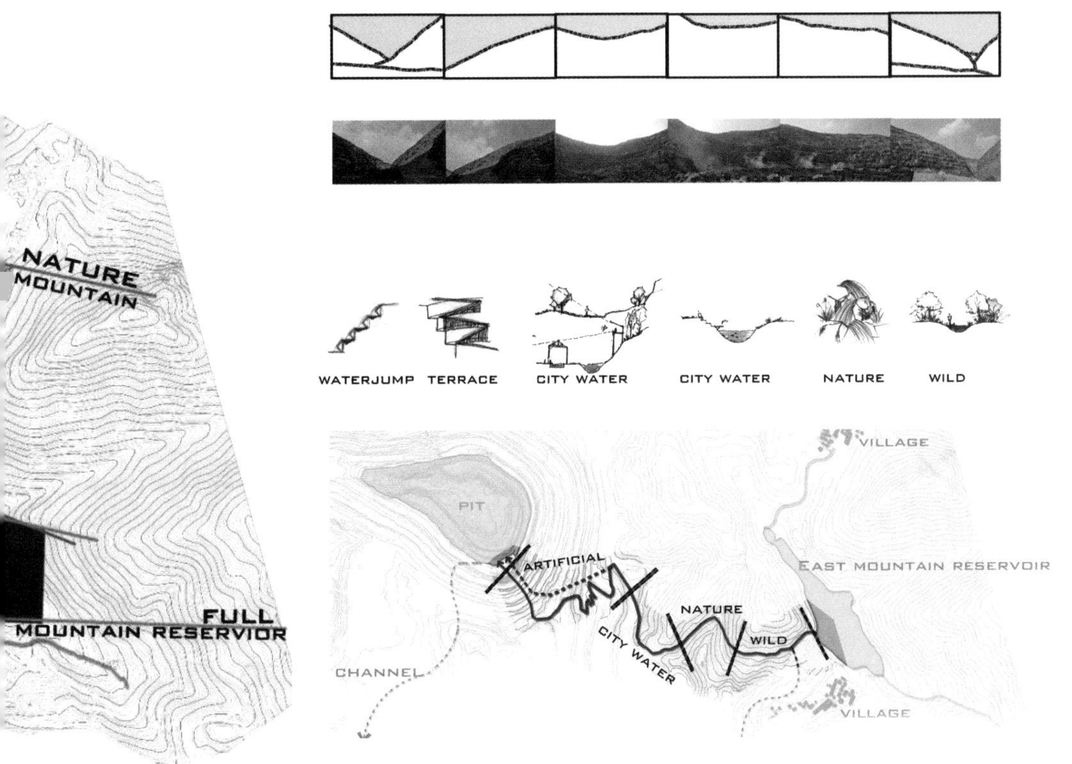

03 总体设计 General Design

总体水系设计

The Water Systerm Design Plan

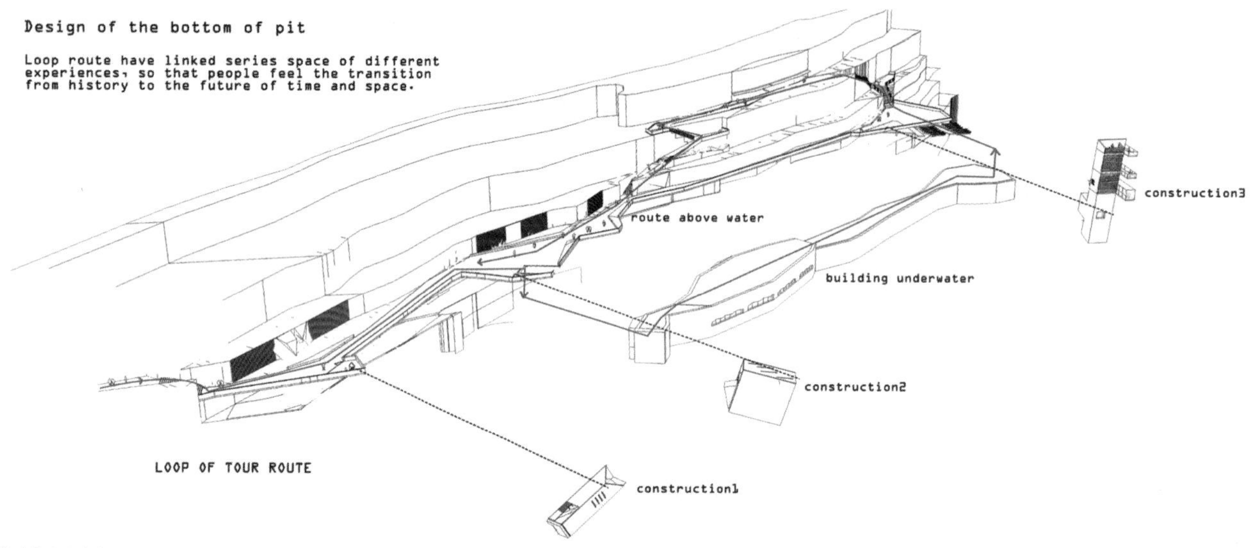

总体路线设计

The Trails Design Plan

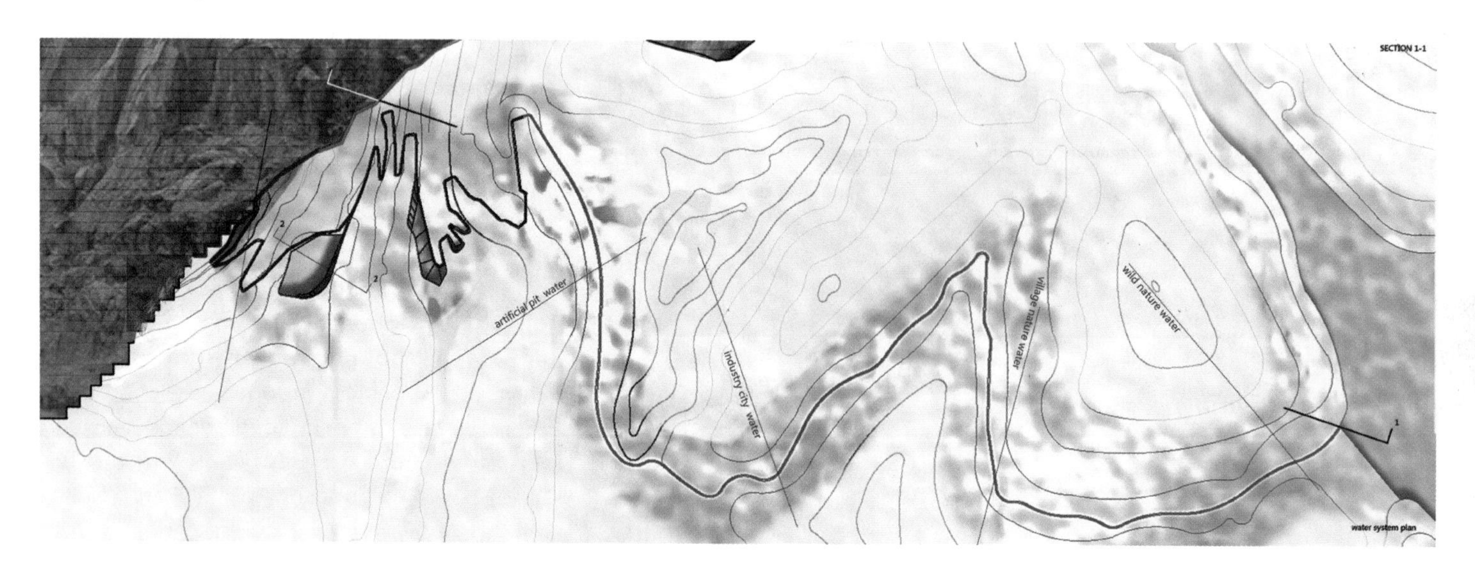

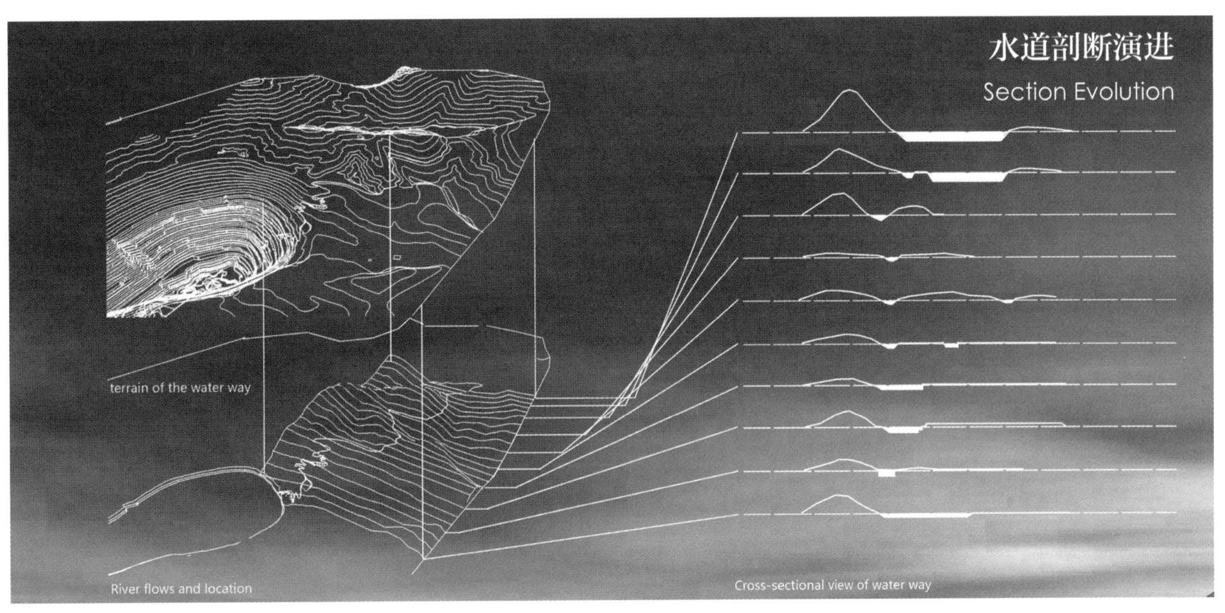

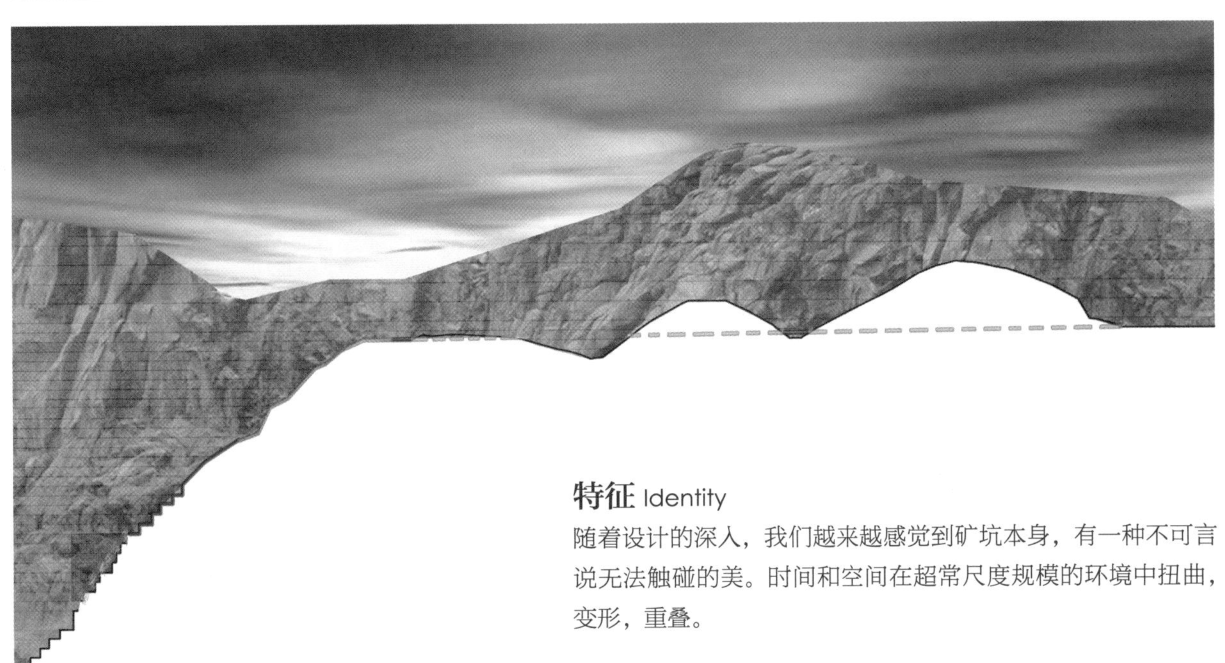

特征 Identity

随着设计的深入，我们越来越感觉到矿坑本身，有一种不可言说无法触碰的美。时间和空间在超常尺度规模的环境中扭曲，变形，重叠。

04 详细设计 Detailed Design

平面设计

The Design Plan

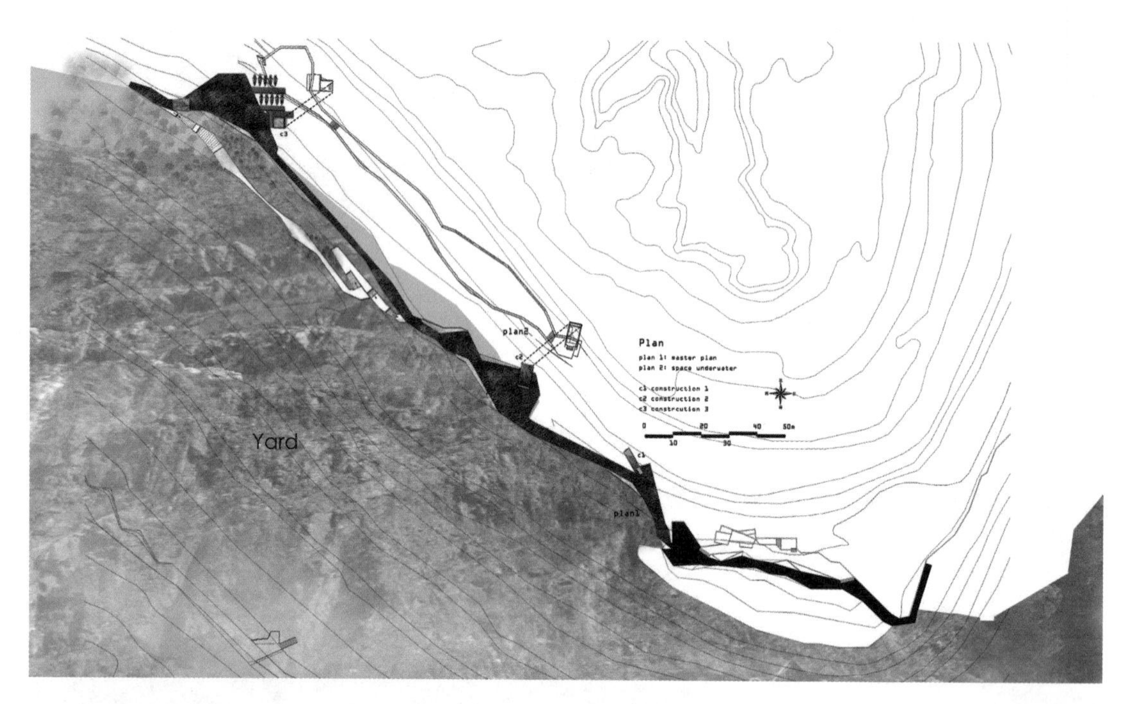

透视图

Perspective

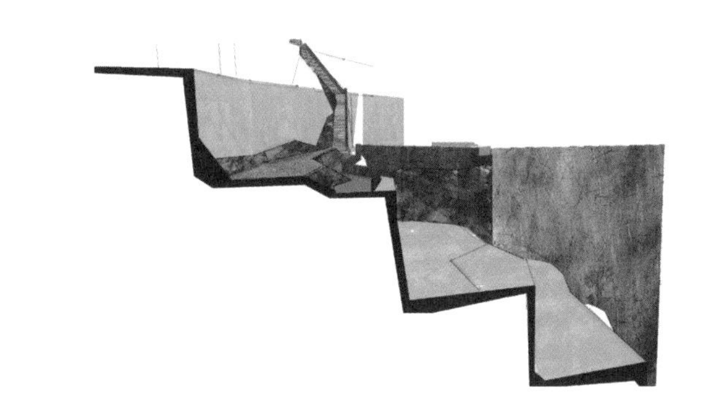

活动展示

Activity Display In Section Underwater

构筑物
Main Construction

活动展示 Activity Display

人们从第二个构筑物，如图所示，下到进入水下建筑部分，开始水下的游览路线，了解矿坑从过去到未来的景观转变过程。

水下游览终点：一个小型的展厅，展示对矿坑的改造过程，然后人们从小型的构筑口到塔台底部，上升，最终回到水面上。

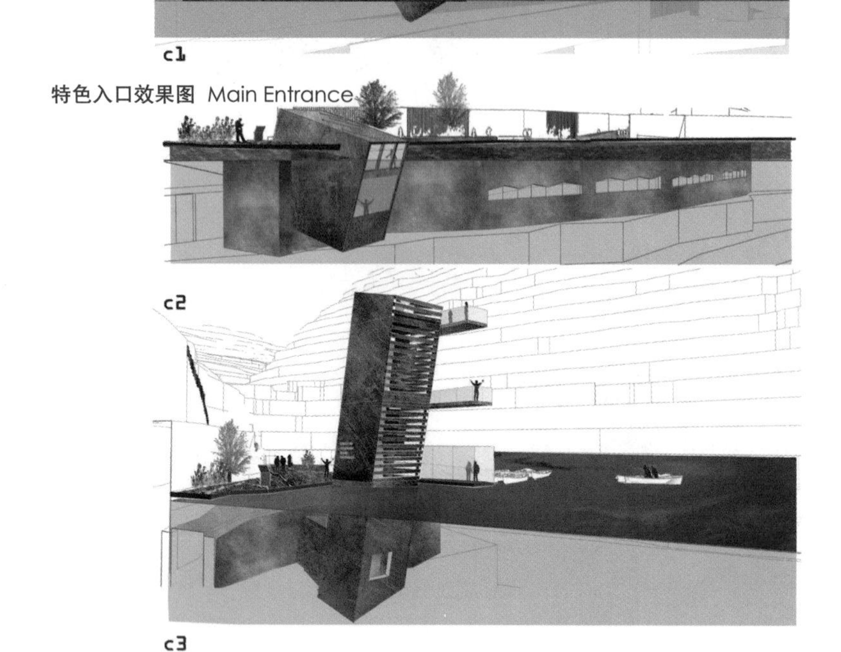

特色入口效果图 Main Entrance

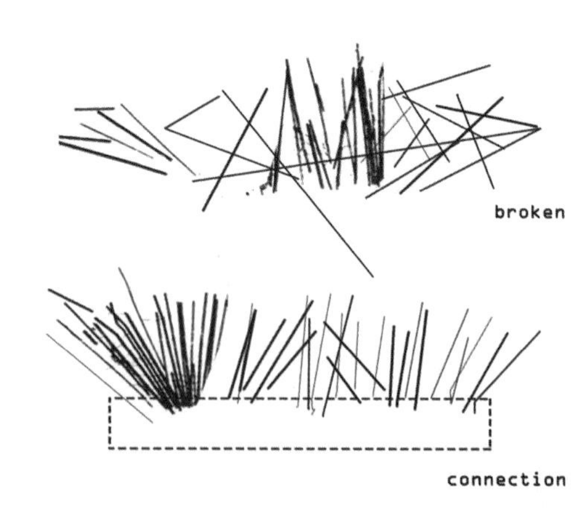

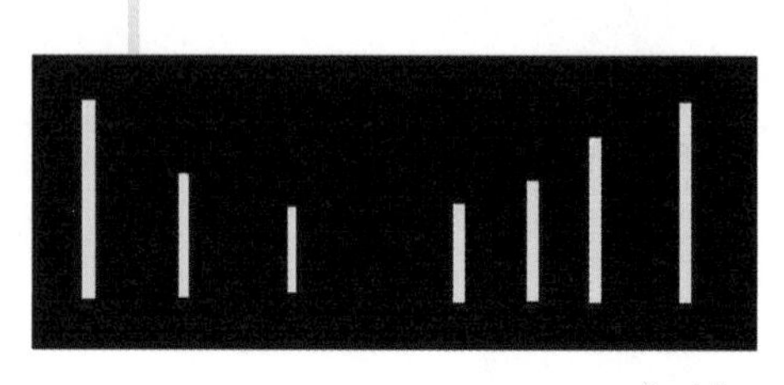

直梯设计——核心构筑物

Elevatorthe Main Structure

因为安全快速下降的需要，我们设计了电梯。电梯进深 40m，外观 70m 高。

Because of security the need of rapid decline, we design the elevator. The elevator depth 40 m, facade 70 m high.

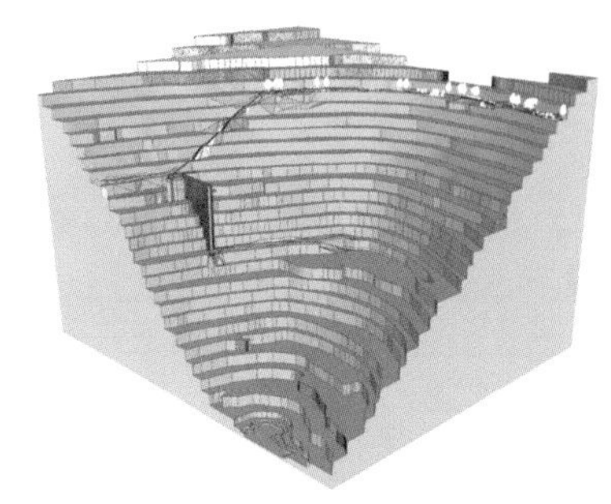

矿业新生，森林重现

The Rebirth of Mining Industry & Forest

秦越 王晓雨 张艳杰
Qing Yue/Wang Xiaoyu/Zhang Yanjie

初到矿区沿着蜿蜒的步行道上山，入口标志塔体现了工业景观的厚重感，入口轴线布置是从人文历史记忆到自然的转换空间，进一步接近矿区我们感受到的是它的震撼，惊叹矿产工人的神奇力量。沿着步行道下行，我们看到的是矿坑的台层特征和工业印记，感慨这座自然的山体中人类留下的伟大力量，在矿坑坑底抬头望天，远看峡谷地带，体验到坑壁近、中、远不同层次的景观变化。在整个矿区的游览中，我们经过了心理从初识——探寻——深究——收获的变化过程。

When we arrived Daye mine, we are strike by its landscape and amazed at mineral workers' magic power. On tour throughout the mine, we go through the mental from acquaintance - explore - get to the bottom - Changes harvesting process.
By studying historical data of the survey, we know that once there is a vibrant green temperate rainforest, and then with the mine's open pit mining, has now become Asia's first sinkhole, great changes occurred in the mine landforms.

秦越
Qing Yue

王晓雨
Wang Xiaoyu

张艳杰
Zhang Yanjie

01 场地现状 Site Analysis

区位分析

District Analysis

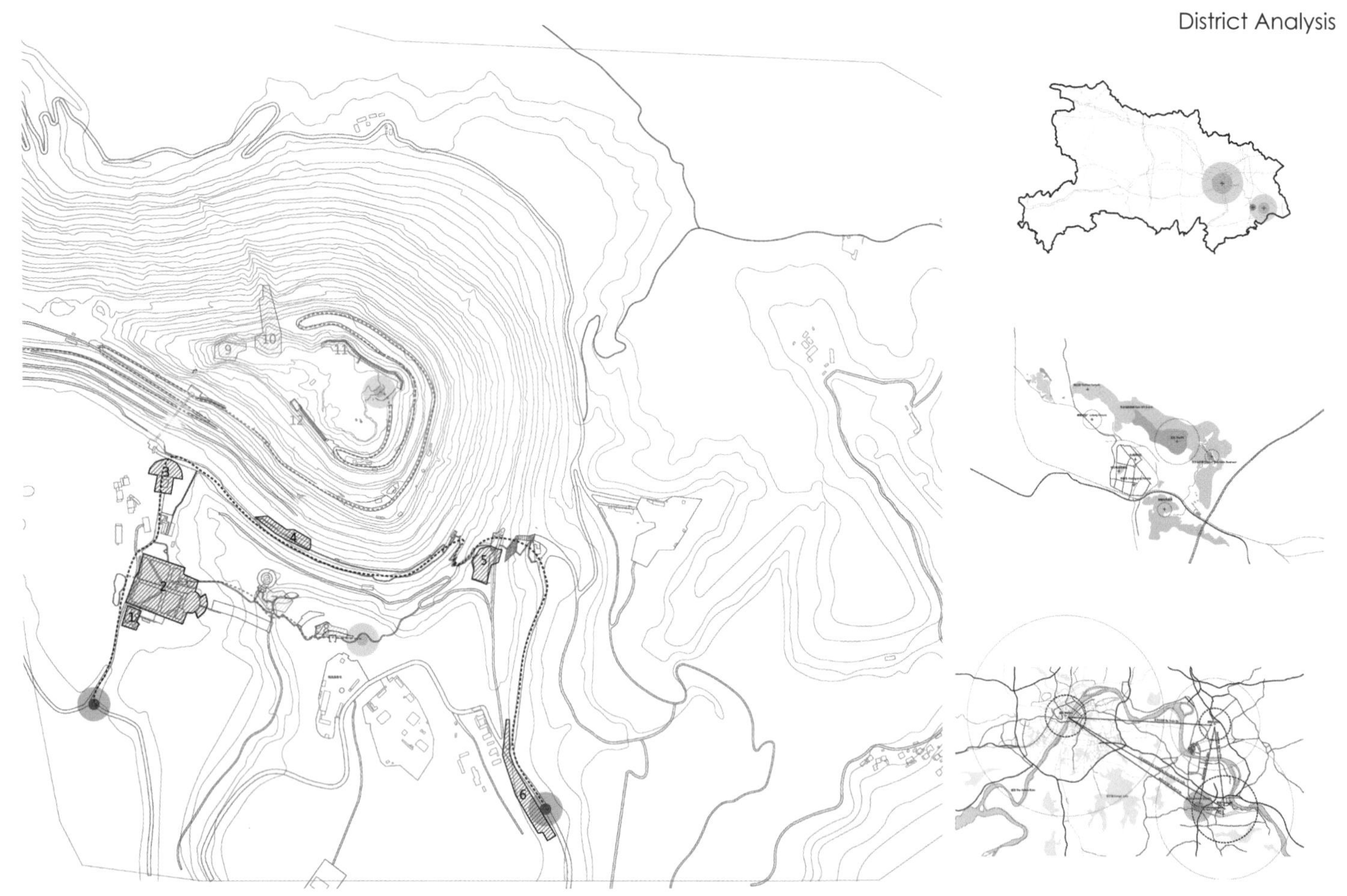

场地感知

Perception of the Site

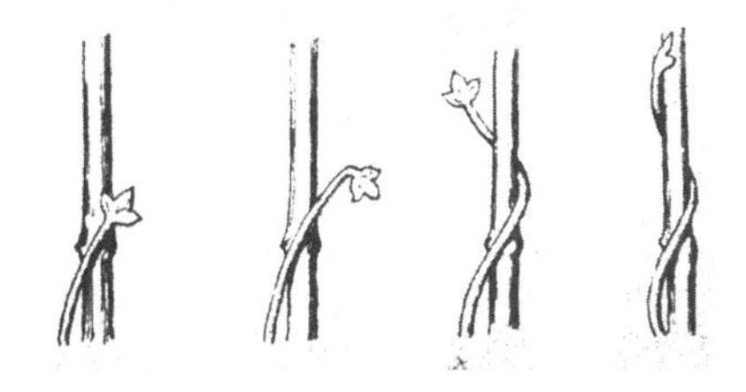

攀援植物的螺旋生长——植物的向上生长运动，对光产生反应。

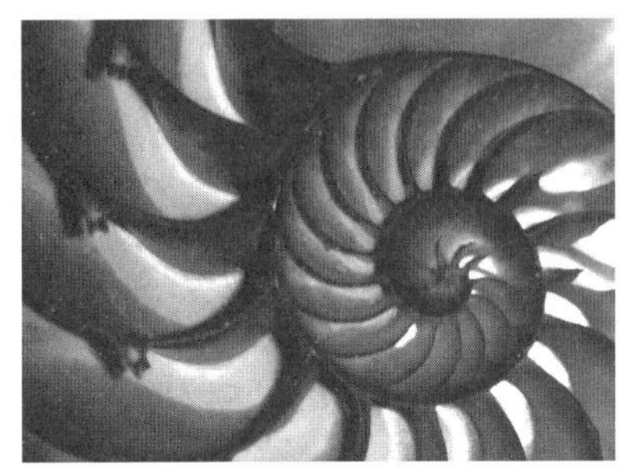

叶子的螺旋生长——为了达到最大量的暴露和最小量的叠生，争取光照和水分。

Nautilus pompilius 鹦鹉螺

叶子的螺旋生长——为了达到最大量的暴露和最小量的叠生，争取光照和水分。

Sempervivum 长生草属

设计概念

Concep

森林景观的再现 Rebuild The "Forest" Landscape

通过对历史资料的调查学习，我们知道曾经这里是绿意盎然充满生机的温带雨林，然后随着对矿山的露天开采，如今已经成为亚洲第一天坑，矿区地貌发生了巨大的变化。经过百年多的开采，绿色的森林已经变为矿石或钢材以建设城市。在长距离的下坑道路中，我们从震惊场地的伟大开始慢慢疲劳，进而期待在坑底区域有一片绿洲。

综上，在设计中我们试图通过森林与场地的对比展现场地的特质，并且满足人们游览时内心的需求。由此衍生出整体设计概念：森林景观的再现。

By studying historical data of the survey, we know that once there is a vibrant green temperate rainforest, and then with the mine's open pit mining, has now become Asia's first sinkhole, great changes occurred in the mine landforms.After more than a century of mining, forest has become ore or steel to build the city.In the distances of the pit road, we from the shock of the great venues outside and then slowly began to look tired in the bottom of the pit region oasis.

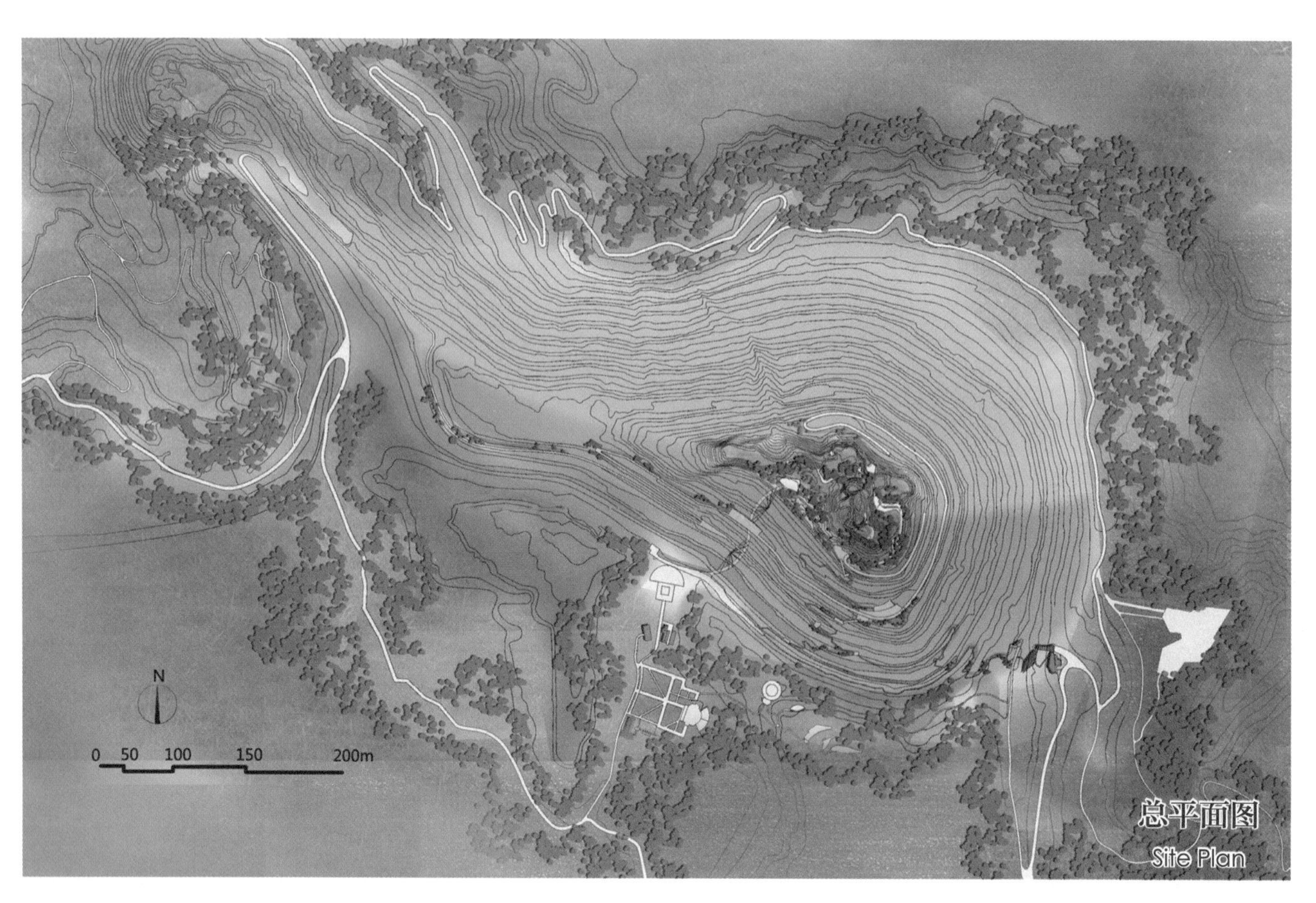

总平面图
Site Plan

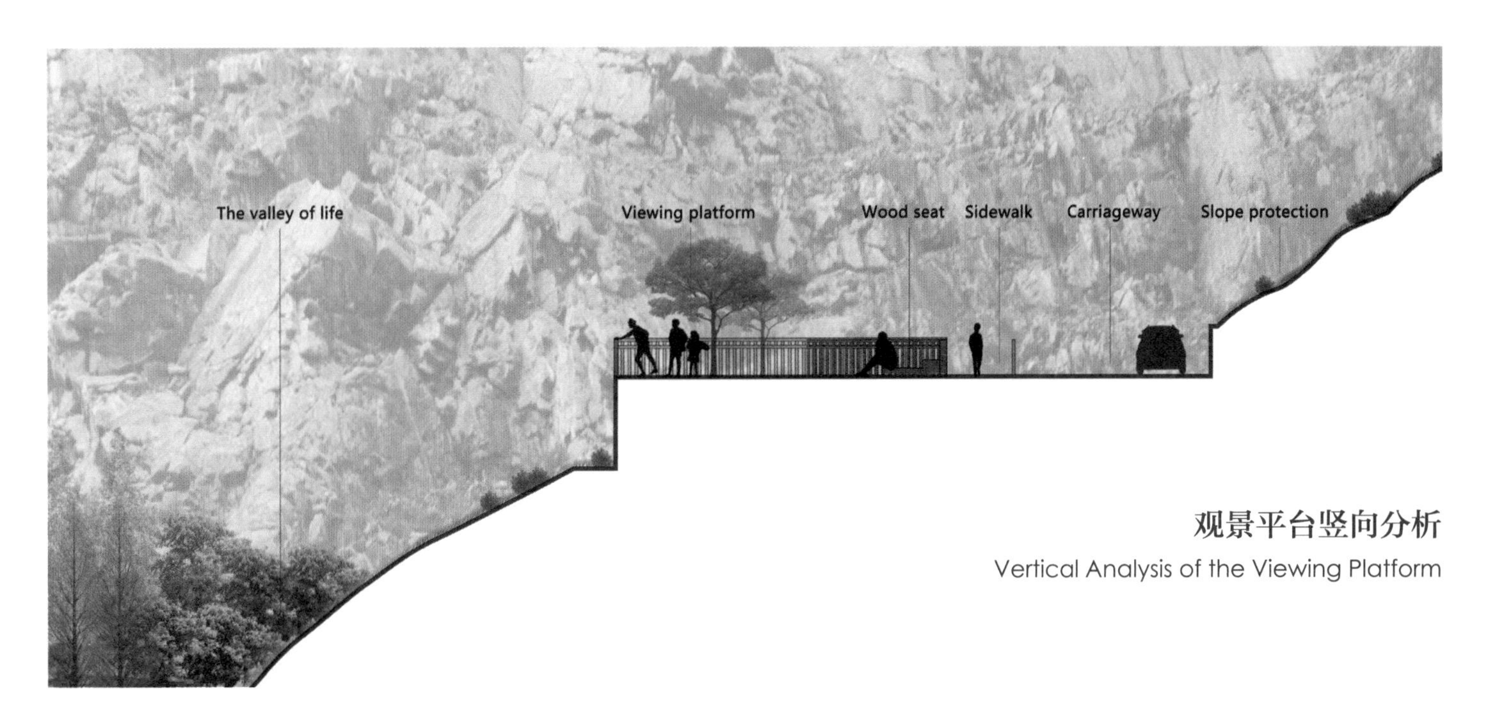

观景平台竖向分析

Vertical Analysis of the Viewing Platform

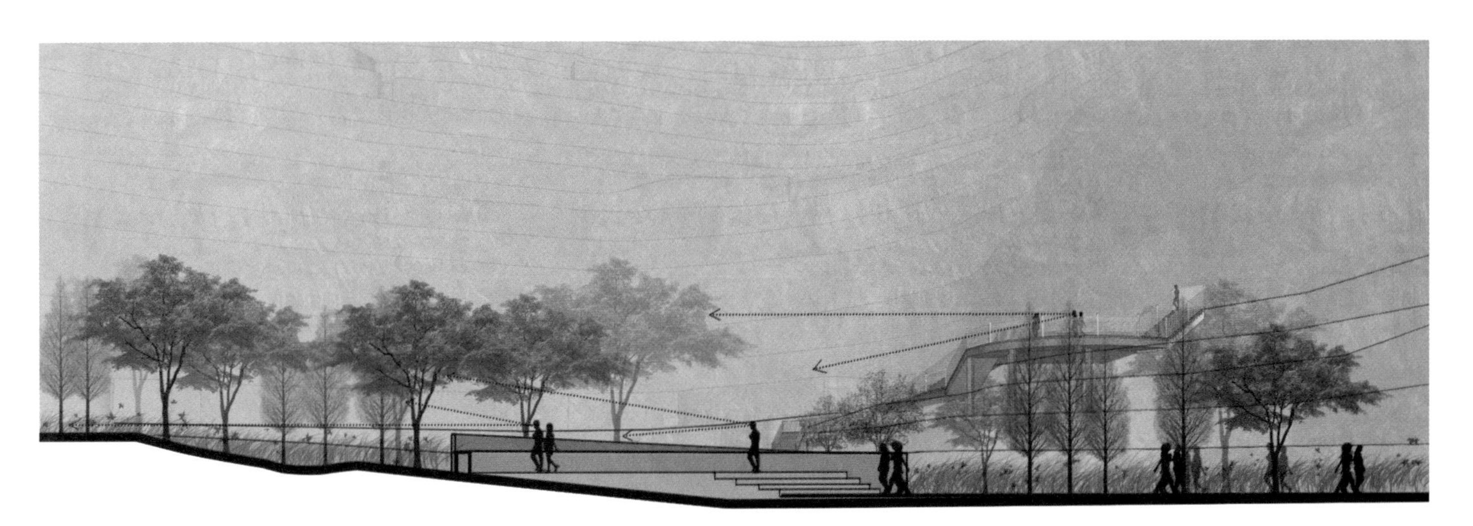

广场景观剖面图

Square Landscape Section

人视点效果图

Perspective

矿坑平面图 Master Plan

洼地初期是无水状态，由于自然汇水的作用，形成自然的水塘，促进周围植物的生长。植物种类逐渐丰富，开始出现灌木，植物量也不断增加。同时能吸引一些小型动物。经过若干年，通过人工以及自然的双重作用，逐渐形成复层群落系统，一个微型生态系统。鸟类以及其他小型动物更容易在此处聚集。

平面图
Plan

剖面图

0 ~ 5 年，地被草花为主，对覆盖的土壤进行修复；

5 ~ 10 年，结合人工干预方式可增加乔木种植，并保证其长势；

10 ~ 20 年，基本能形成复层群落结构人工森林景观。

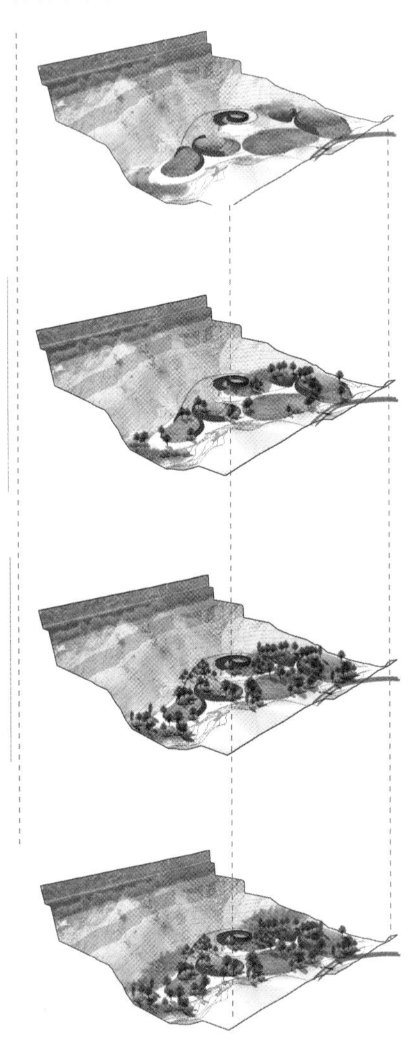

矿坑景观剖面图

Quarry Landscape Section

透视图

Perspective

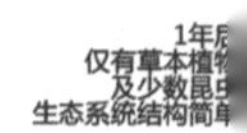

One year
Herbaceo
and little c

植物
plant

动物
animal

5年后
有丰富草本植物种类及简单灌木
有昆虫及栖息于灌草的鸟类
生态系统较为简单

Five years later

Many herbaceous plants and some shrubs

Insects and birds

20年后
初步形成乔—灌—草复层结构
动植物种类增加
初步形成坑底生态系统

Twenty years later

Herbaceous plants,shrubs,trees

Many animals

100年后
形成层次丰富，种类繁多的复层生态结构
动植物种类丰富
坑底生态系统完全形成

One hundred years later

Rich plant structure and animal species

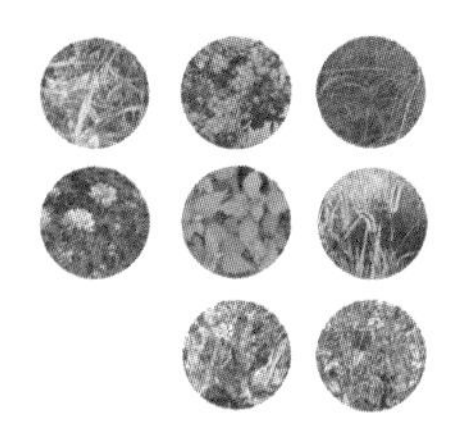

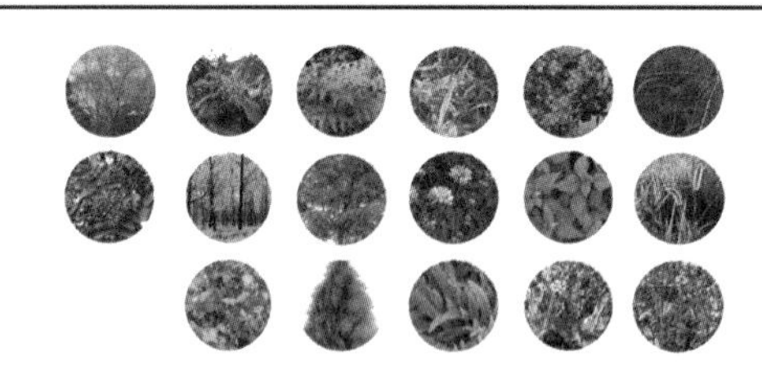

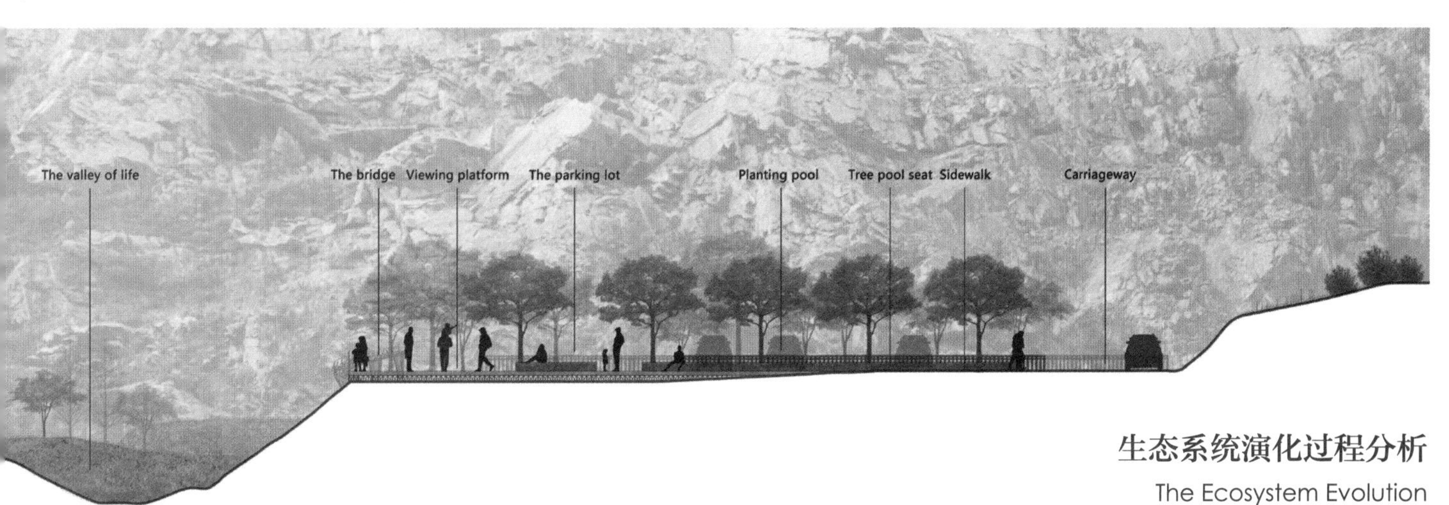

生态系统演化过程分析

The Ecosystem Evolution

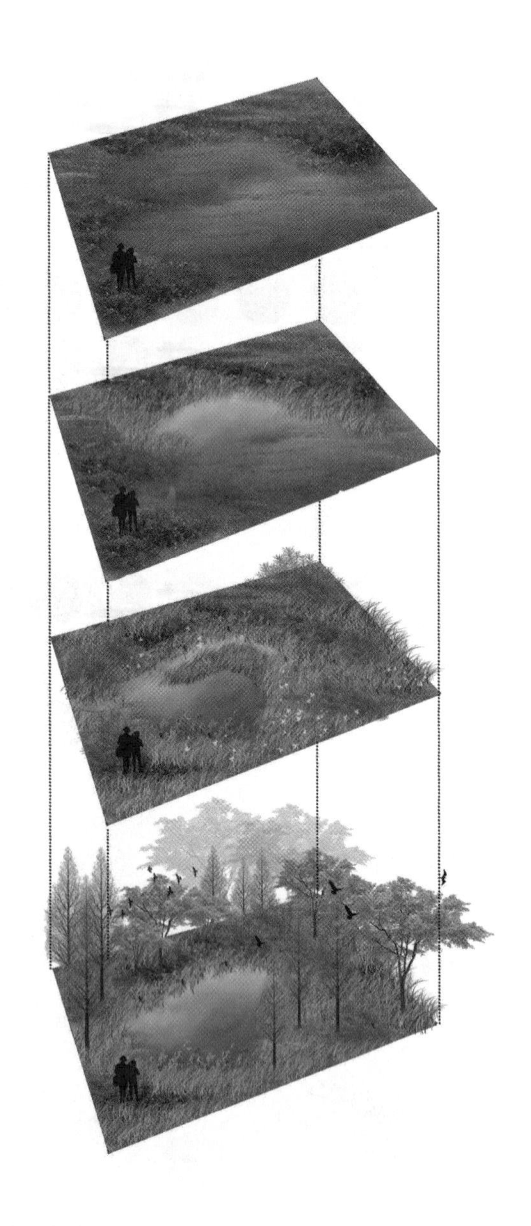

1年后
仅有草本植物
及少数昆虫
生态系统结构简单
5年后
有丰富草本植物种类及简单灌木
有昆虫及栖息于灌草的鸟类
生态系统较为简单

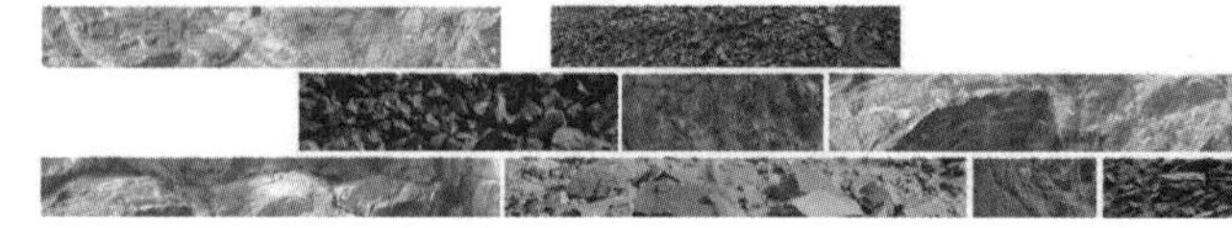

生态系统演化过程分析

The Ecosystem Evolution

20年后
初步形成乔–灌–草复层结构
动植物种类增加
初步形成坑底生态系统

100年后
形成层次丰富，种类繁多的复层生态结构
动植物种类丰富
坑底生态系统完全形成

粉色流水线

PINK ASSEMBLY LINE

荣南　华锐　王笑时
Rong Nan/Hua Rui/Wang Xiaoshi

对于废弃工厂的改造，有任其荒废改造成公园的，也有改造成游乐园的，也有改造成博物馆的。总结下来就是基本都保留了原有建筑的外形，进行内部功能的置入。但是，曾经仅仅依靠改造成公园、游乐园或者博物馆的案例，现在都面临着严峻的挑战，资金无法获得后续支持，直接导致场地最终被推平转卖。

因此，我们想到了能够提供持续动力的因素，引入创意工作室、商业这些新的功能，并且也兼具博物馆、展览和户外休闲的功能。

For the innovation,many cases give the idea such us change it into a park and treat the abandoned factory as sculptures. Summary comes down almost all good cases has retain the original building and introduce new function.But almost all these cases are facing severe chanllenges now that not enough finance to keep the situation.

So, we want to find an anwser can provide continuous power.At last we introduce the creative studio and commerce to import more people and museum and exhibition to tell people the knowledge.

01 场地现状 Site Analysis

参观序列 The Visit Sequence

根据场地调研的结果，场地原有的生产路径分成三条，按照三组建筑从矿山开采到最终制成出厂。这是工厂运行时候的原料输送路径，如今工厂停产倒闭了，但这一条线路包含着原有场地的历史，应当至少保留一条供人参观。

According to the results of the site investigation, the production route is divided into three.It's composed by three buildings from mining to the end of the route.It's how the factory produce the cement before,but now the factory has closed down as a result the route was abandoned.For us,this route contains the history of the site,so at last we should retain one line for visitors.

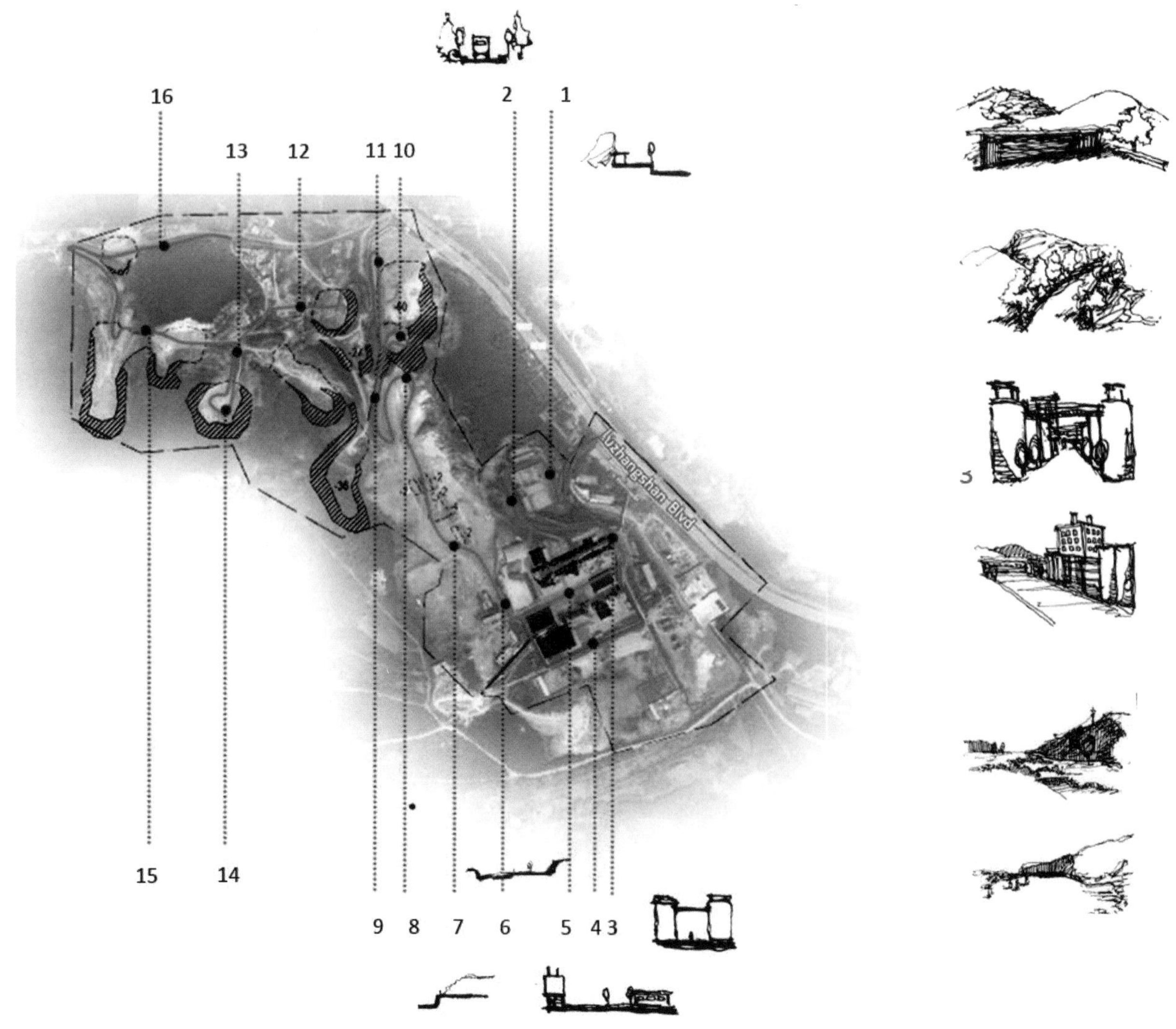

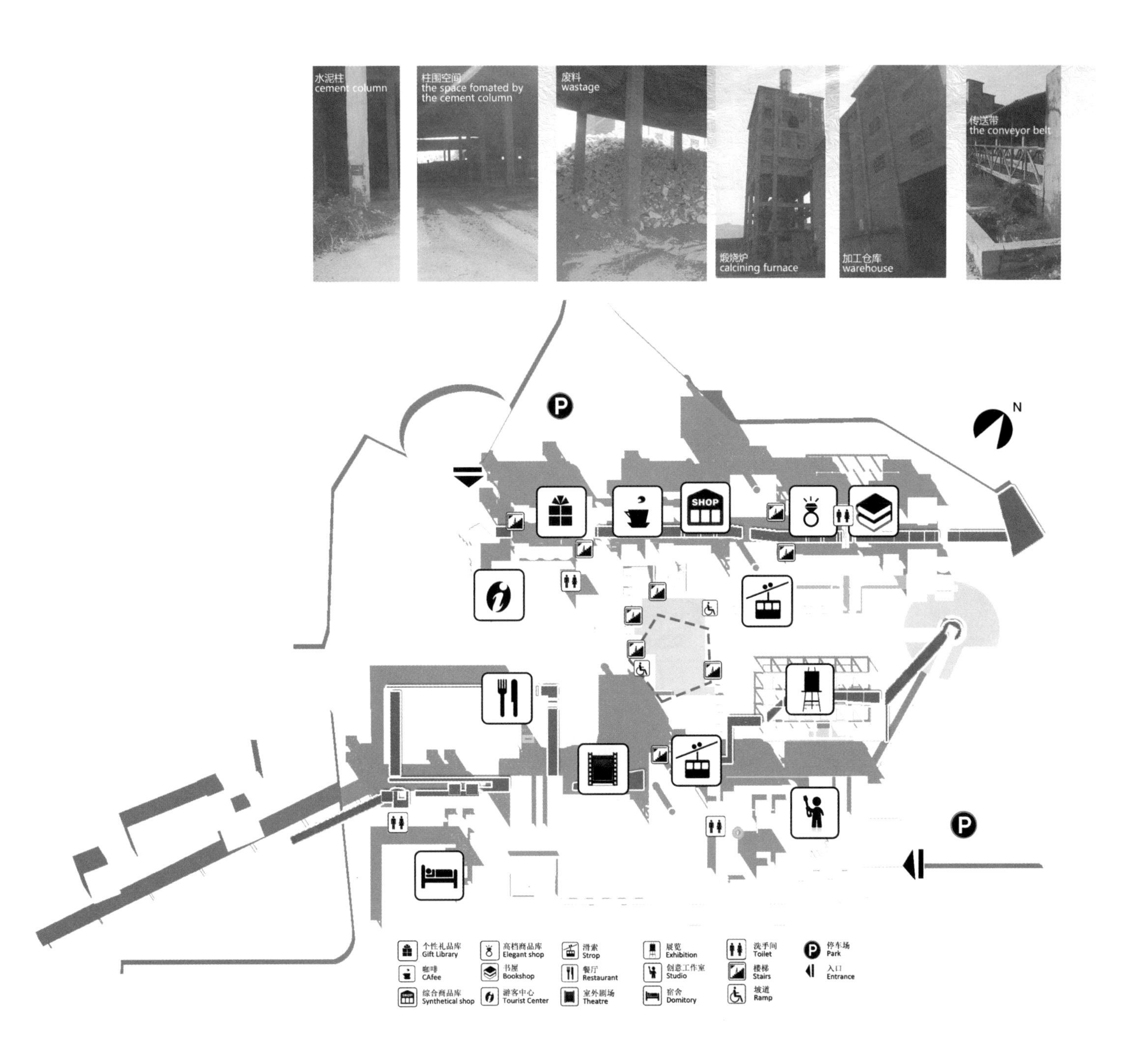
水泥柱
cement column
柱围空间
the space fomated by the cement column
废料
wastage
煅烧炉
calcining furnace
加工仓库
warehouse
传送带
the conveyor belt
SHOP
N
个性礼品库
Gift Library
咖啡
CAfee
综合商品库
Synthetical shop
高档商品库
Elegant shop
书屋
Bookshop
游客中心
Tourist Center
滑索
Strop
餐厅
Restaurant
室外剧场
Theatre
展览
Exhibition
创意工作室
Studio
宿舍
Domitory
洗手间
Toilet
楼梯
Stairs
坡道
Ramp
停车场
Park
入口
Entrance

02 理念构思 Concept

概念 Concept

此外，我们想到了能够提供持续动力的因素，引入创意工作室、商业这些新的功能，并且也兼具博物馆、展览和户外休闲的功能。

we want to find an anwser can provide continuous power.At last we introduce the creative studio and commerce to import more people and museum and exhibition to tell people the knowledge.

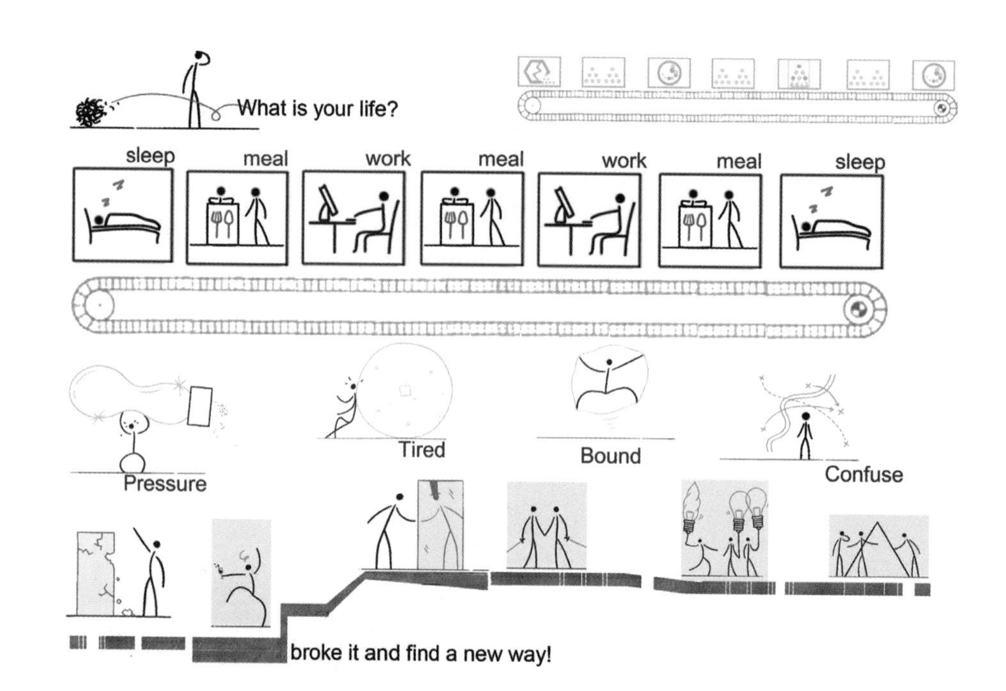

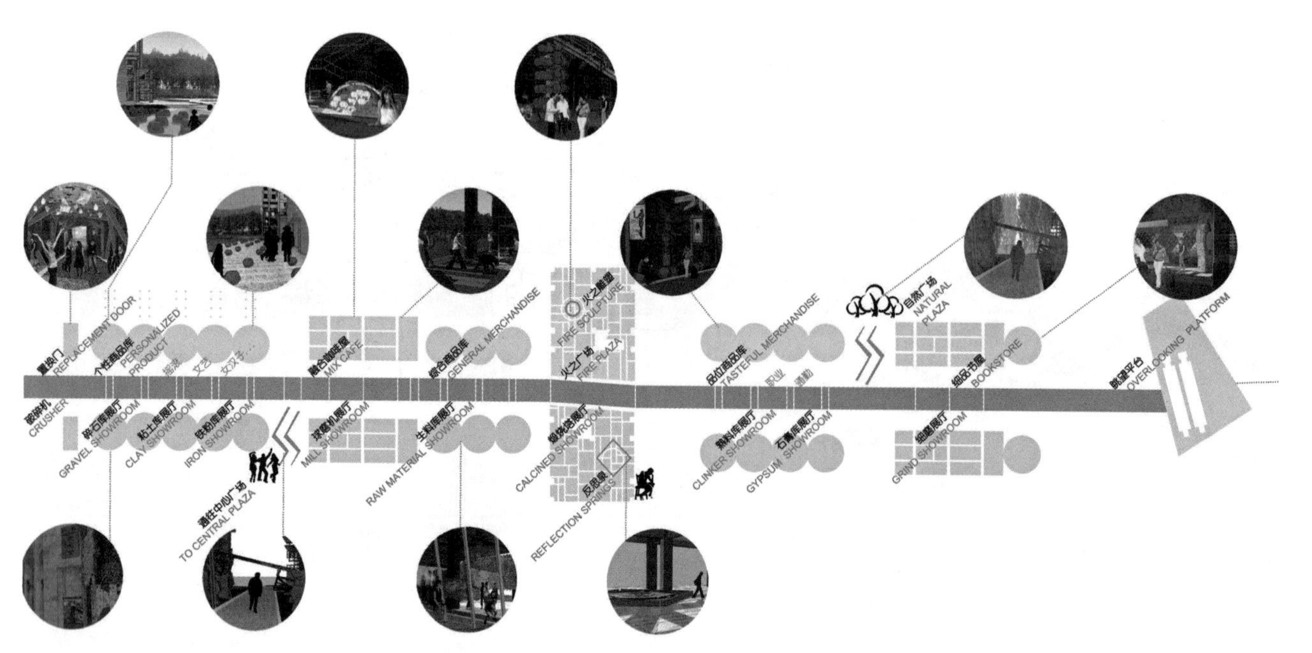

总平面图 PLAN

后工业究竟带给了我们什么？带给了我们效率、统一。但同时也开始缩减差异。我们与在生产线上被输送的原料，有着某种程度的相似。后工业带给我们的印记，其实就是这种人被物化的悲哀。而设计的目的就在于新置功能的同时，带给人们有关后工业的反思。

What does the post industrial bring us? The mark post industrial has brought for us was the materialization.And the aim of the design was bring rethinking for the people while place new function.

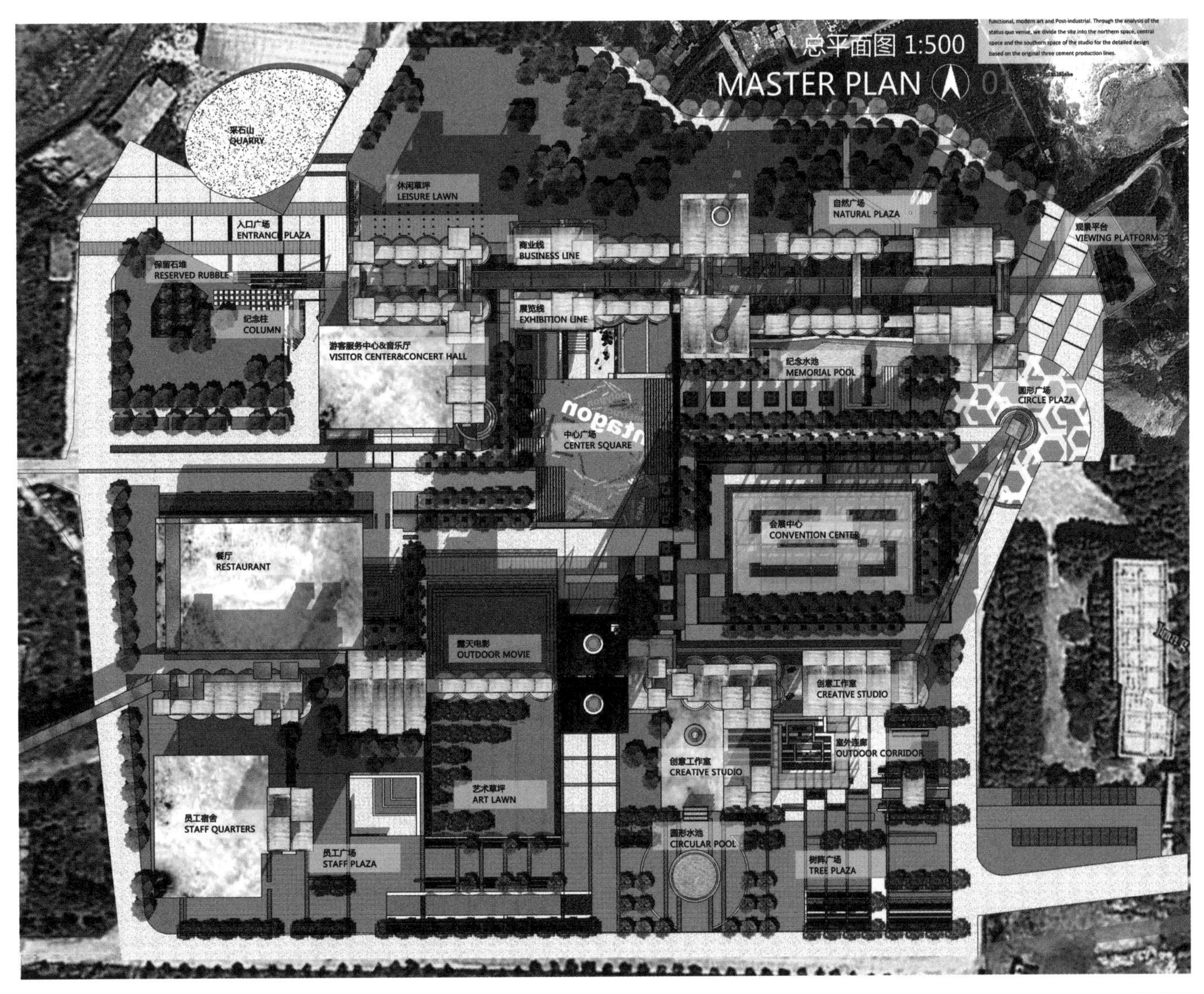

整体鸟瞰
Bird's View

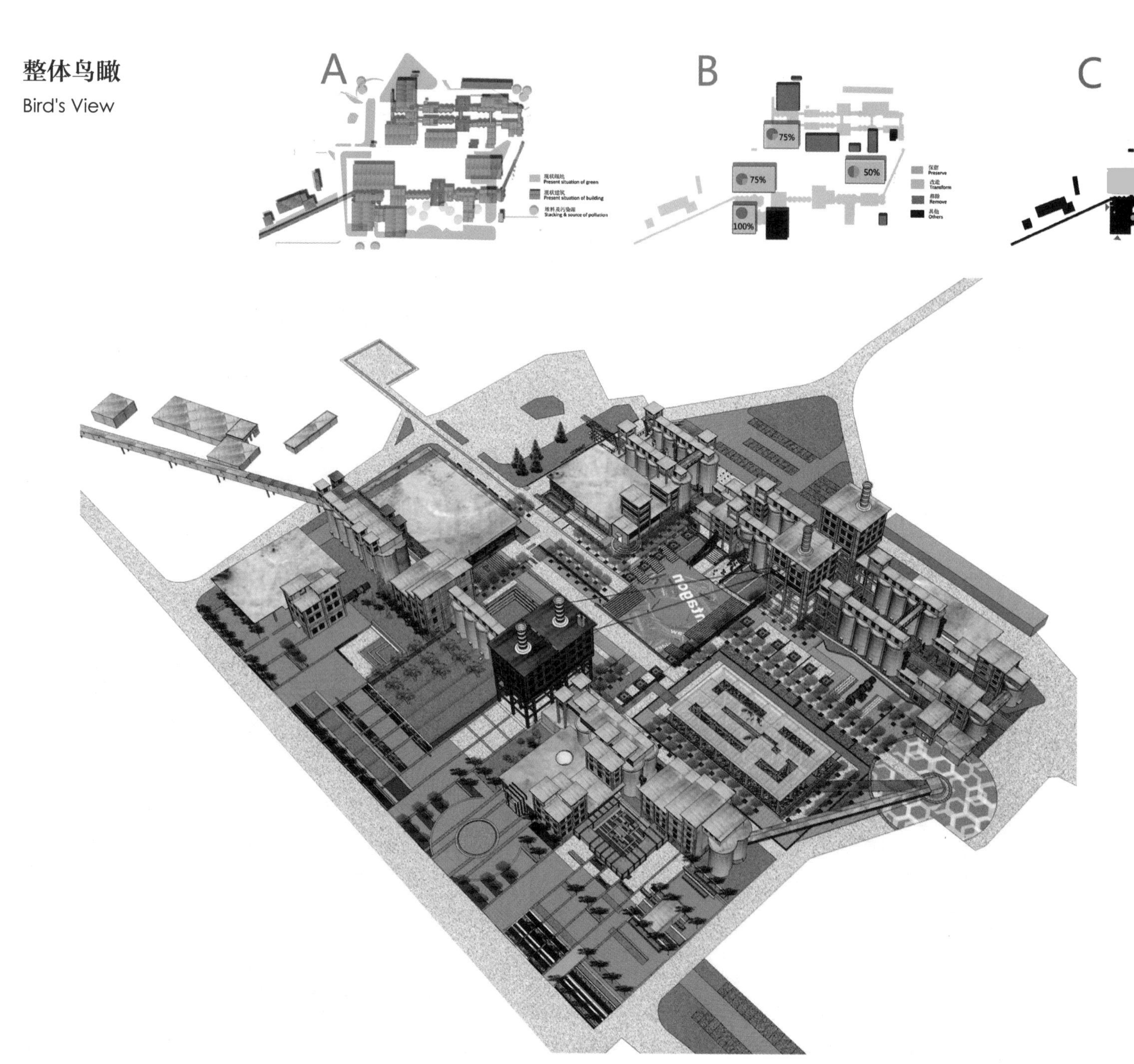

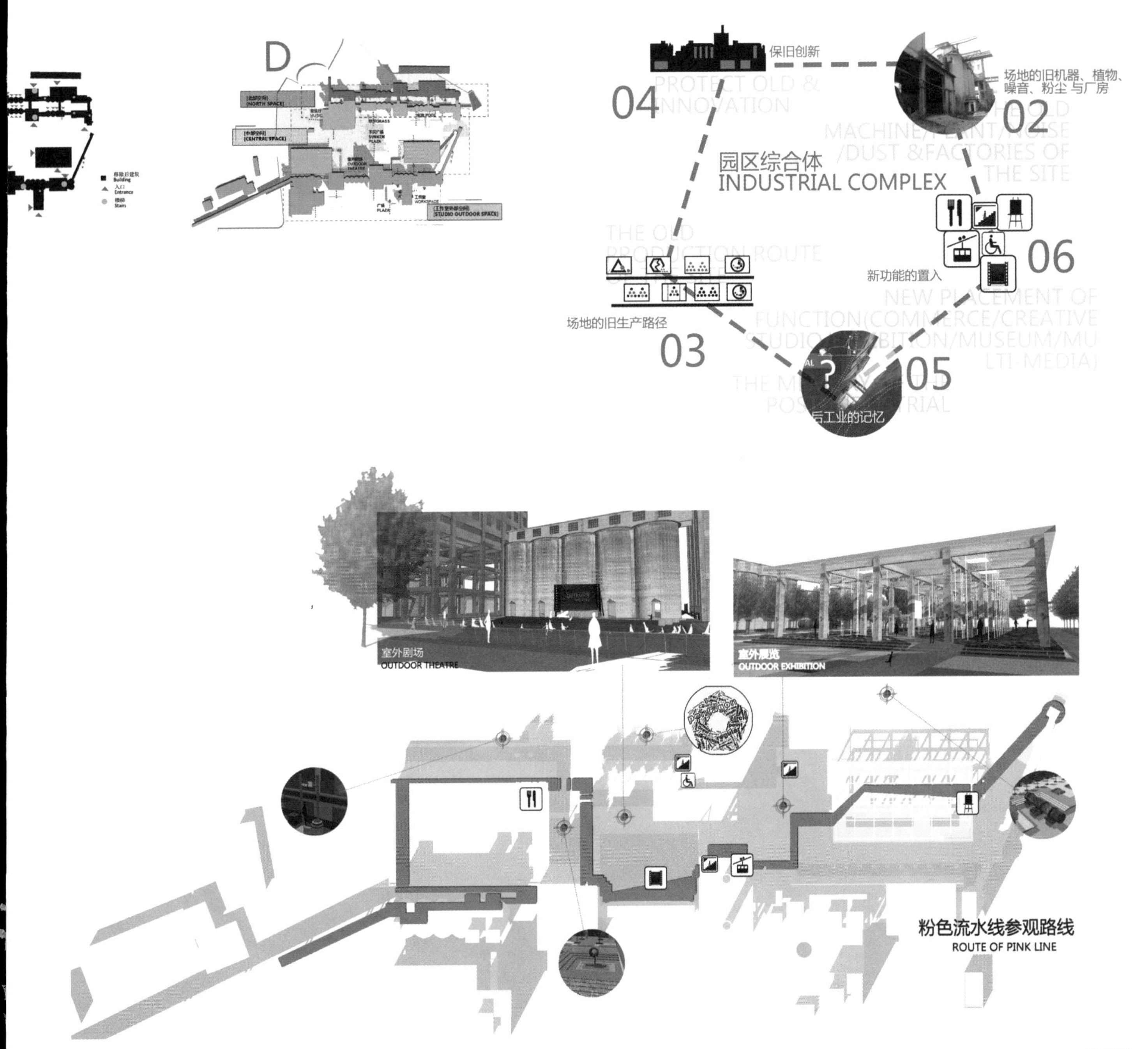
保旧创新
PROTECT OLD & INNOVATION
04
场地的旧机器、植物、噪音、粉尘与厂房
THE OLD MACHINE/PLANT/NOISE /DUST &FACTORIES OF THE SITE
02
园区综合体
INDUSTRIAL COMPLEX
THE OLD PRODUCTION ROUTE OF THE SITE
场地的旧生产路径
03
新功能的置入
NEW PLACEMENT OF FUNCTION(COMMERCE/CREATIVE STUDIO/EXHIBITION/MUSEUM/MULTI-MEDIA)
06
后工业的记忆
THE MEMORY OF THE POST INDUSTRIAL
05
室外剧场
OUTDOOR THEATRE
室外展览
OUTDOOR EXHIBITION
粉色流水线参观路线
ROUTE OF PINK LINE

场地剖面设计
Site Section

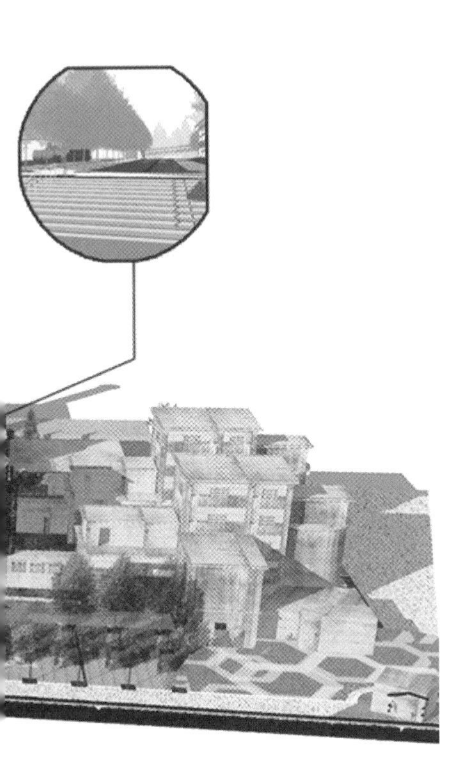

下沉广场 Sunken Plaza

这个部分是整个园区的中心，因此又称作中心广场，主要承担着集散、聚会等多种功能。在这里参观者可以很好地欣赏到周边建筑提供给人的印象，也可以从这里进入建筑内部的参观路线并选择滑索运动。

This part is the center of this site, therefore is called as the center square,which is mainly play a important role in the function of distributed,panty etc. In there visitors can not only have a good time to appreciate the environment of surrounding buildings,but also select the strop movement by get into the route in buildings from there.

透视效果图

Perspective

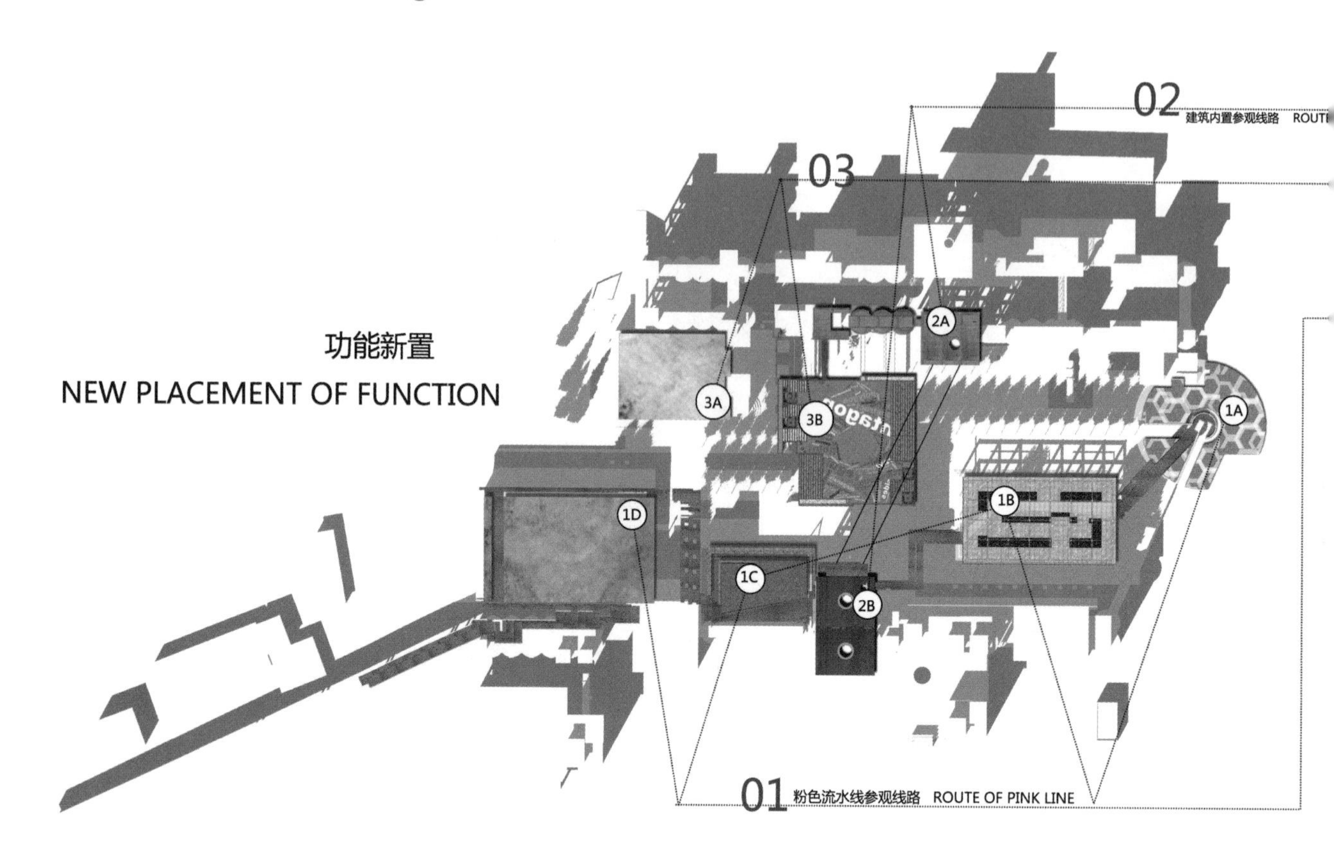

功能新置
NEW PLACEMENT OF FUNCTION
02
建筑内置参观线路 ROUTE
03
2A
3A
3B
1A
1B
1D
1C
2B
01
粉色流水线参观线路 ROUTE OF PINK LINE

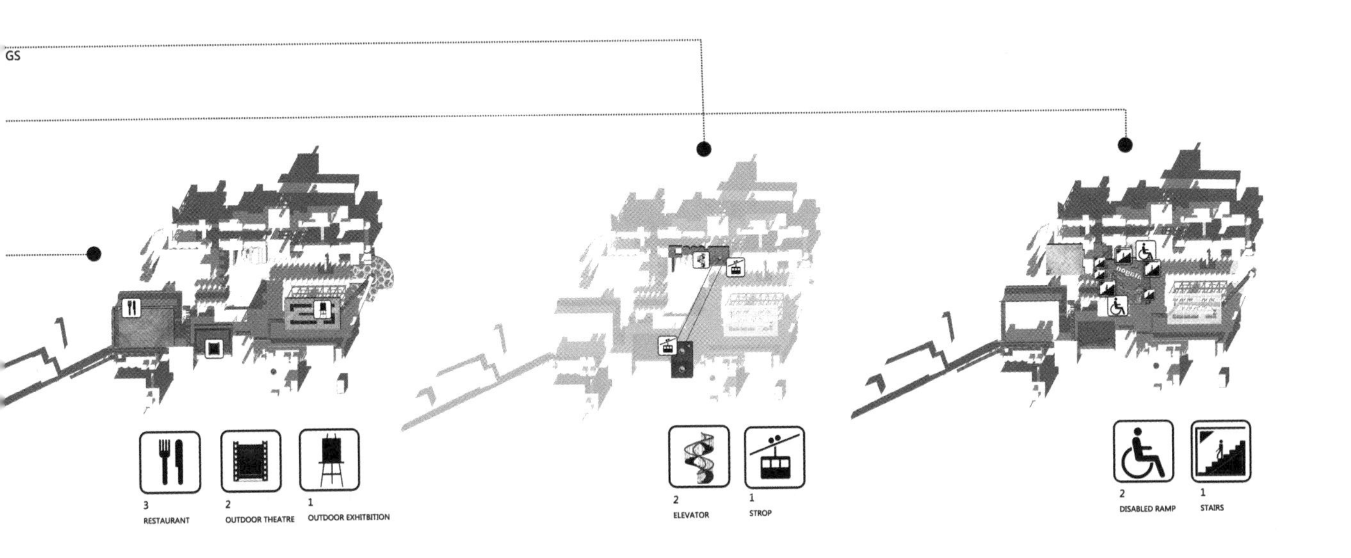

室外展览 Outdoor Exhibition

对建筑南侧流水线的堆料库进行改造，保留其廊柱的构架，去掉原有的薄板墙体，变成一个开放式的展厅。在粉色流水线的两侧对成品水泥的种类进行一些图片展示，外围则是定期对园区创意工作室的作品进行展览，增进文化更新的可持续性。

The tansformation of this cement warehousespace in the flow line was retaining the framework of it and remove the wall before,then it become a opened exhibition space.On the both sides of the pink line show the species of finished cement,abd the external is to show the production from the creative studio there regularly, so that this site can enhance the sustainability of cultural renewal.

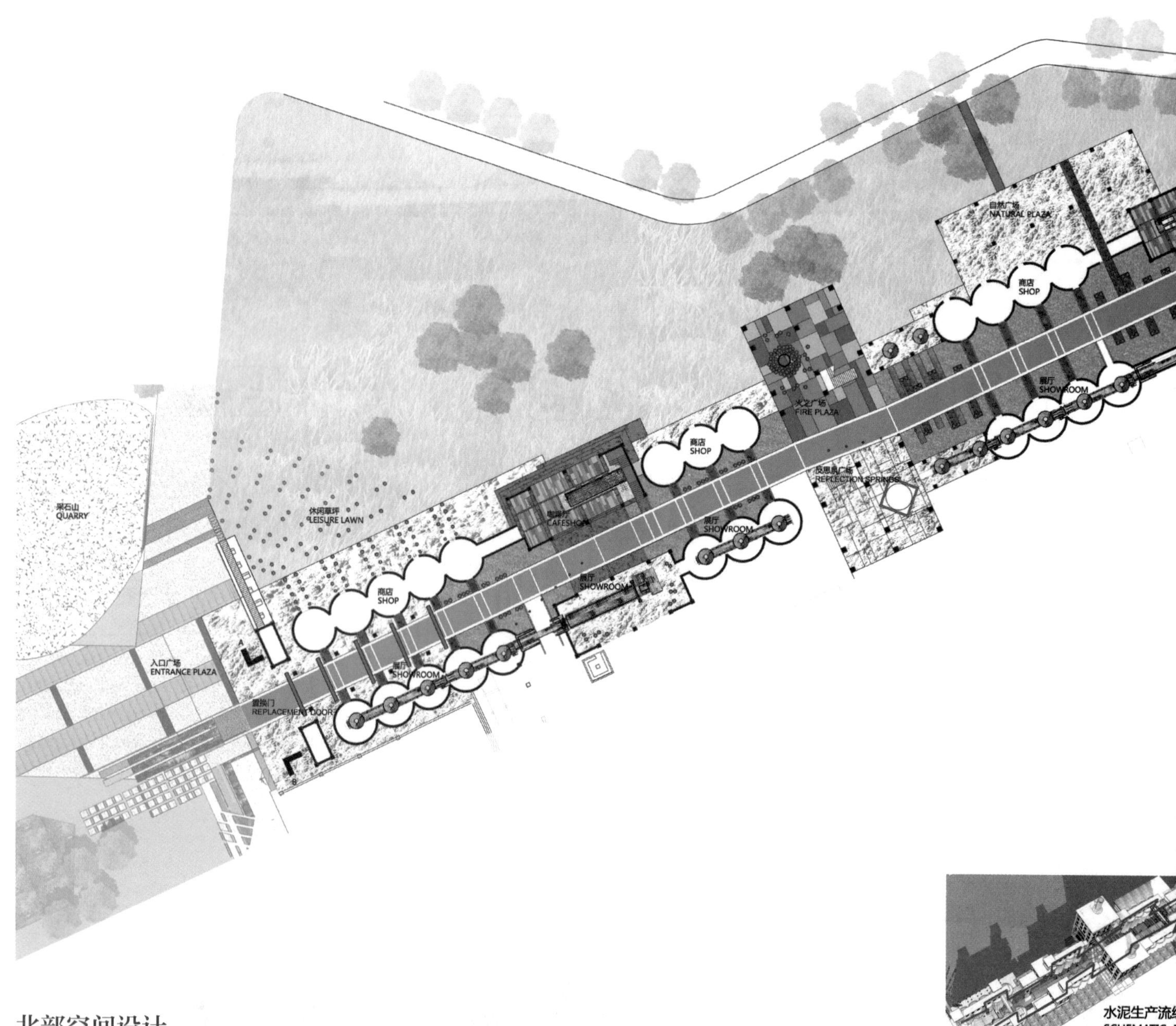

北部空间设计

The Design Of North Space

空间分析与透视

Analysis And Perspective

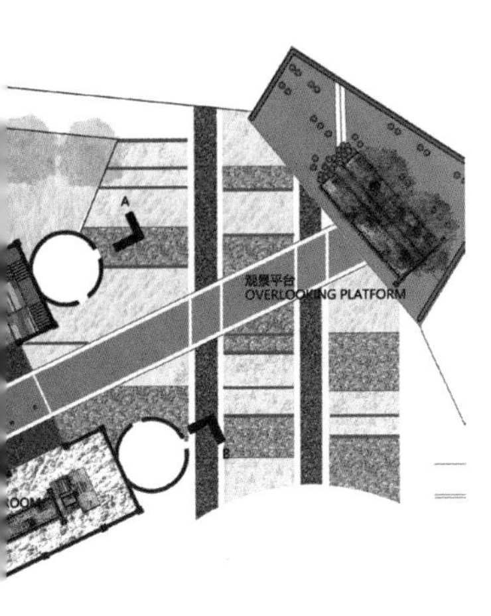

北部空间是两条生产线中的线性空间，设计保留南侧的整体生产线作为水泥工艺流程的展示馆。根据原有生产线的特征，改造北侧的建筑，作为商业用途，同时体现原有特征，使人们在街道空间行走时可以感受两侧的不同和联系，产生与历史的共存感。

Two production lines in northern space is linear space, designed to retain the south side of the whole process of cement production line as the exhibition hall. According to the characteristics of the original production line, the transformation of the north side of the building, for commercial purposes, while reflecting the original features, so that people walking in the street space can feel and touch the different sides, resulting in a sense of the history of coexistence.

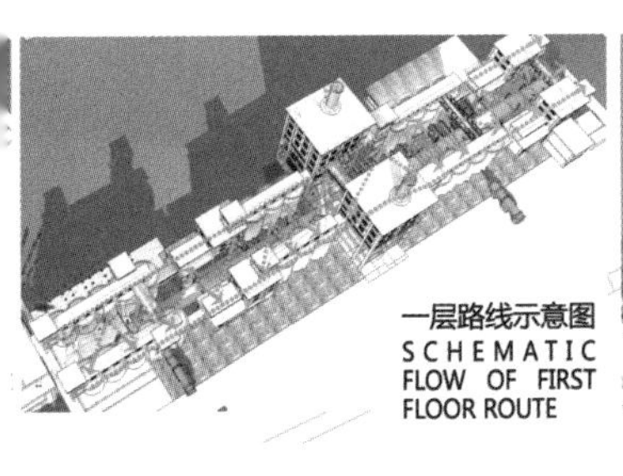

一层路线示意图
SCHEMATIC FLOW OF FIRST FLOOR ROUTE

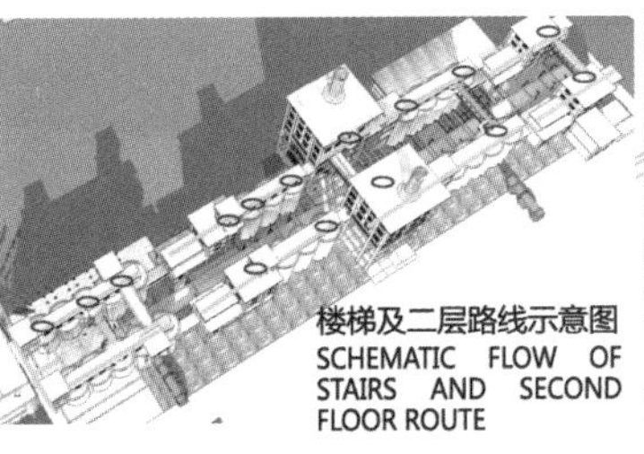

楼梯及二层路线示意图
SCHEMATIC FLOW OF STAIRS AND SECOND FLOOR ROUTE

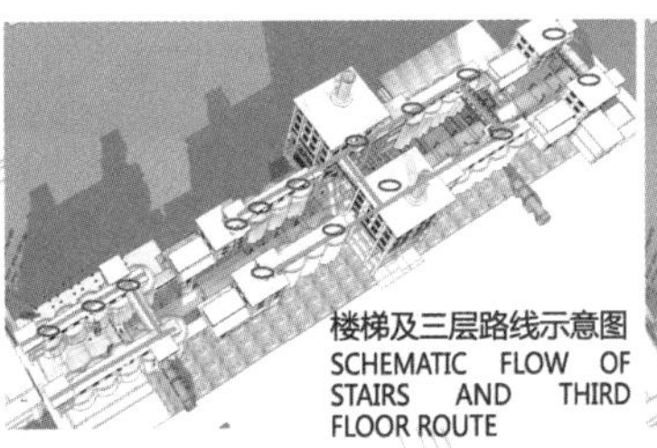

楼梯及三层路线示意图
SCHEMATIC FLOW OF STAIRS AND THIRD FLOOR ROUTE

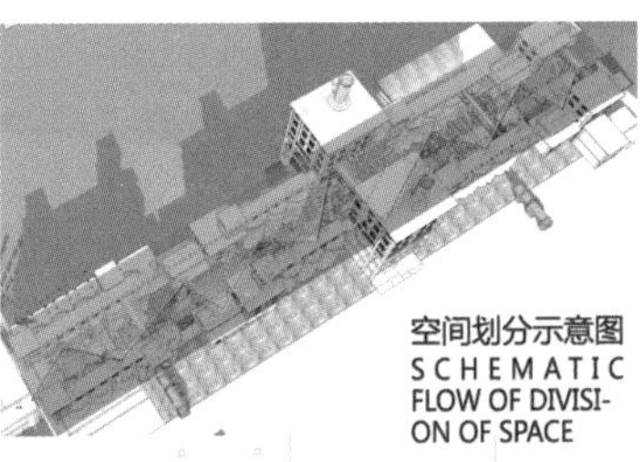

空间划分示意图
SCHEMATIC FLOW OF DIVISION OF SPACE

废弃场地景观设计

LPAM

李颖睿　关学国　周泽林　马严彦
Li Yingrui/Guan Xueguo/Zhou Zelin/Ma Yanyan

大冶铁矿有着悠久的历史，是毛泽东主席生前视察的唯一矿山。随着历史的演变大冶铁矿经历了从官办到官督商办到商办再到国有化的发展历程。1938 年曾被日本占领，给大冶铁矿留下一段不可磨灭的记忆。20 世纪以来大冶铁矿在 50 年间开采矿石约 1.3 亿吨，如今的大冶铁矿已经成为亚洲第一天坑。

通过对历史资料的调查学习，我们知道曾经这里是绿意盎然充满生机的温带雨林，然后随着对矿山的露天开采，如今已经成为亚洲第一天坑，矿区地貌发生了巨大的变化。

Daye Iron Mine has a long history , Chairman Mao Zedong during his lifetime was the only mine inspections. With the evolution of history Daye Iron Mine has gone from government-run Government Commerce to do business office and then to the development process of nationalization. In 1938, Japan was occupied, leaving some indelible memory to Daye Iron. Since the 20th century, Daye Iron ore mined in 50 years about 130 million tons, and now has become the first Asian sinkhole.

By studying historical data of the survey, we know that once there is a vibrant green temperate rainforest, and then with the mine's open pit mining, has now become Asia's first sinkhole, great changes occurred in the mine landforms.

李颖睿
Li Yingrui

关学国
Guan Xueguo

周泽林
Zhou Zelin

马严彦
Ma Yanyan

01 场地现状 Site Analysis

矿坑历史发展认知 History Of The Pit

1. 规划范围的确定

项目的具体规划范围为北侧至铁山区的行政边界，东侧为山体中的公路，南侧是生态山水与城市的边界，西侧为山脊线。

2. 矿坑历史发展认知

公元 226 年，古代开采。1890 年，近代开采。1938 年，日本略采。1952 年，新中国成立后重建开采。2003 年，深部开采，资源枯竭。铁矿保有储量锐减，经营发展的"瓶颈"；经济产值下降，将会造成铁山区人口流失。

3. 场地感知与旅游资源分析

项目基地有丰富的历史特色包括历史名人、历史工业器械、遗留建筑等。有多样的生态特色，诸如裸露的山崖、茂盛的植被、生态斑块、水流痕迹等。有多样的空间特色，像开敞空间、峡谷空间、巨型围合空间、流动空间、密林空间、台地空间等。有多样的肌理特色，如水平与垂直、圆形、波浪形、流动的曲线。有多样的景观场景，有峡谷 + 植被 + 建筑、天坑 + 螺旋形、台地 + 生态斑块、自然山体 + 城市背景等。

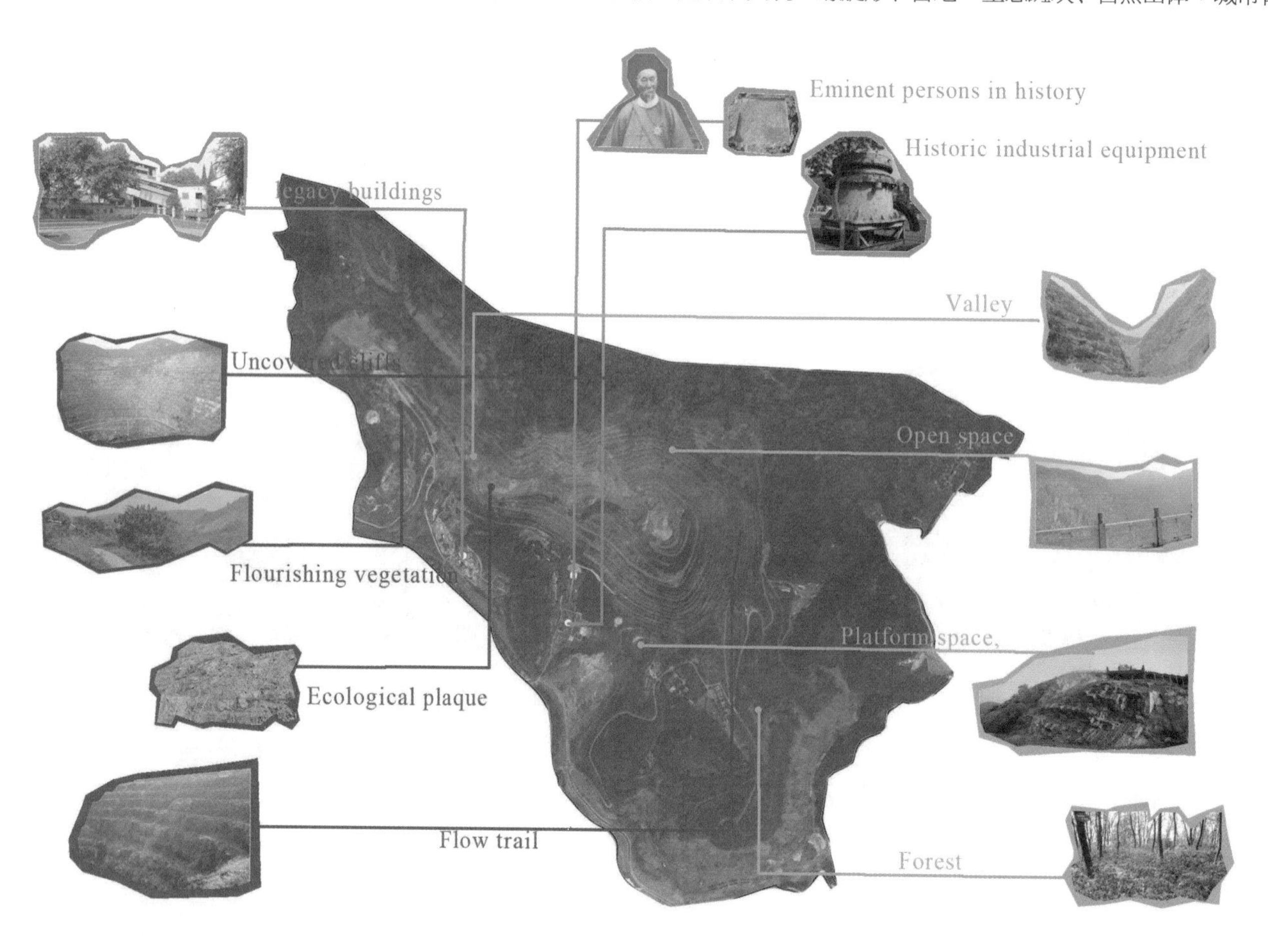

场地感知与旅游资源分析

Sites Perception And Tourism Resources

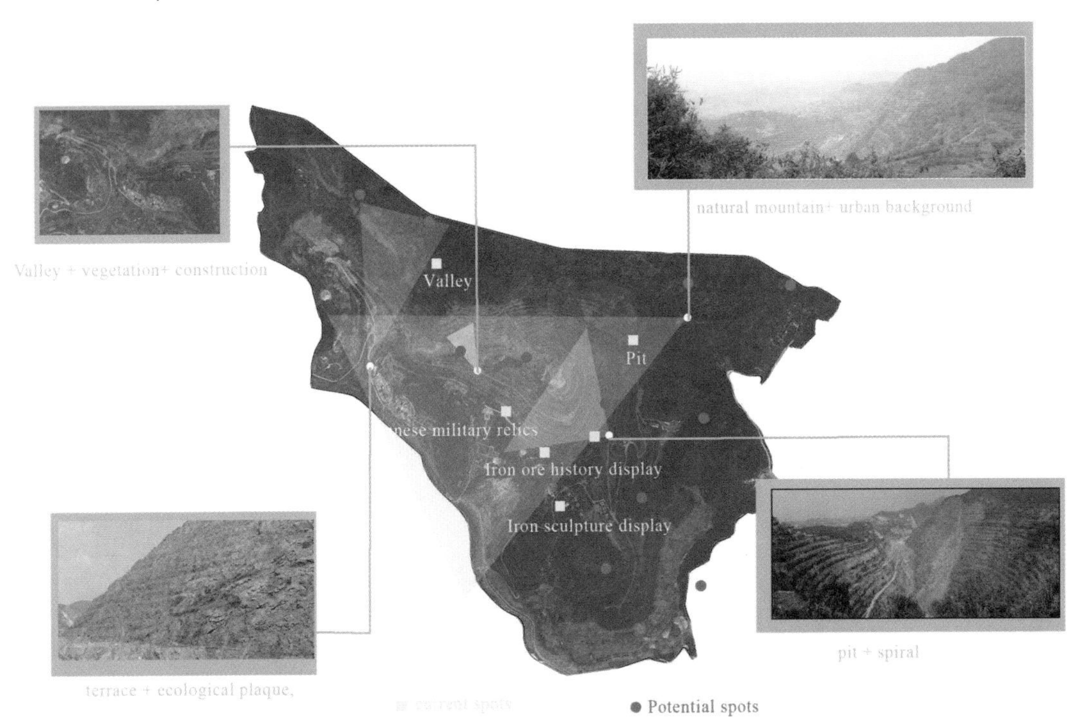

坑体形成过程

The Process Of The Pit

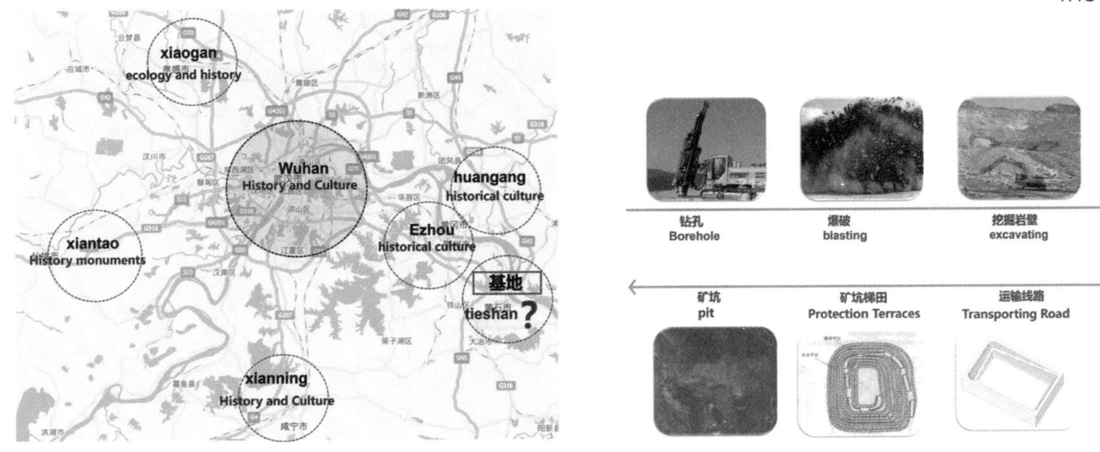

总体规划平面图

Layout Of Master Planning

把空间还给自然还给历史

人们对这块场地干扰很严重，将其由山变成了坑，现在人类以经济为目的的干扰基本结束，是时候将这块场地还给自然，通过设计的手段让游人能看到这里自然的力量和历史留下的痕迹。

Knowing more about the nature and history of the spot

Human made the hill into pit. It is almost finish that interference for the basic purpose of economic. It is time that return nature and history to the spot. Though the design to notice the nature elements and the trail of history.

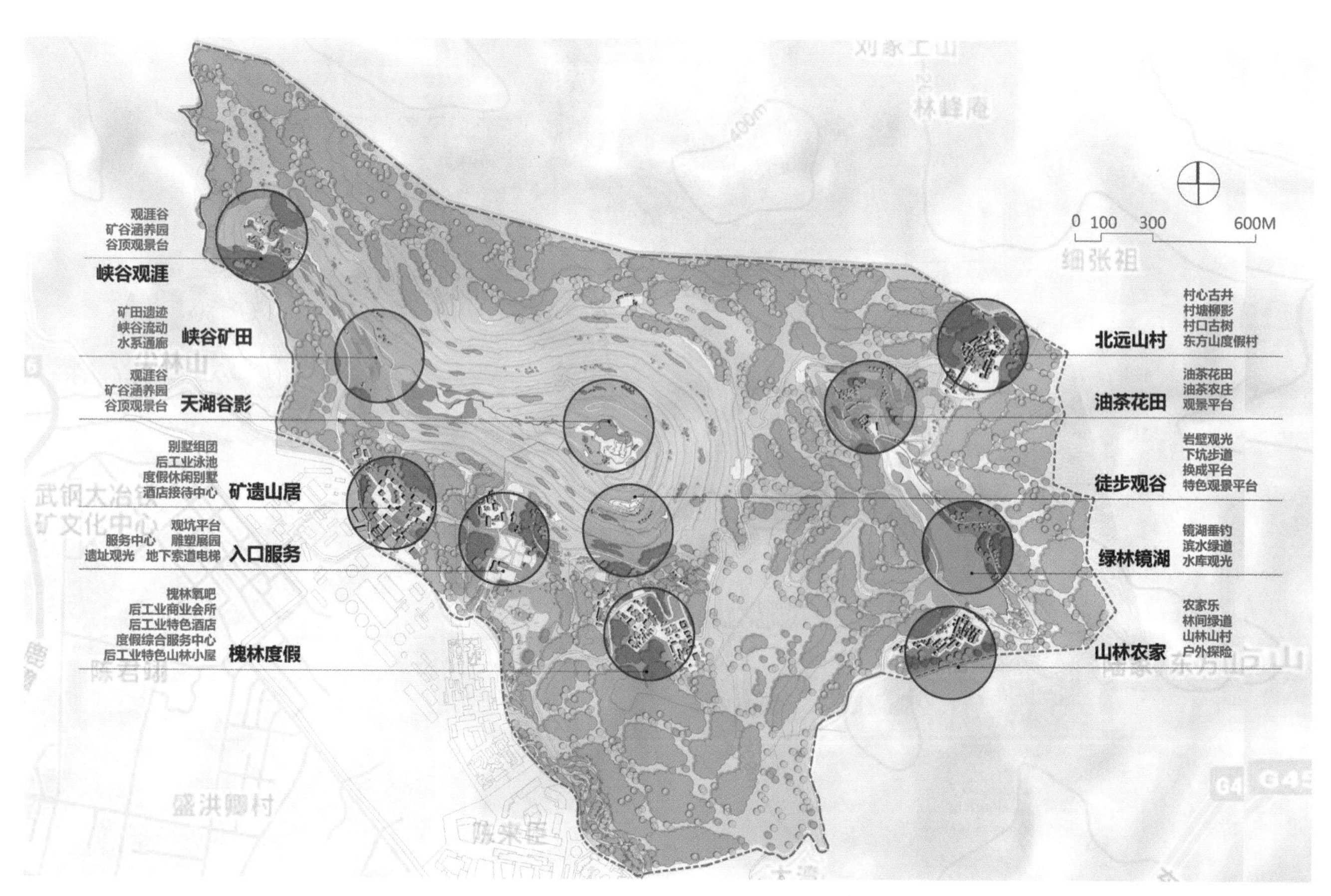

详细规划——坑体剖面结构

Pit's Section Structure

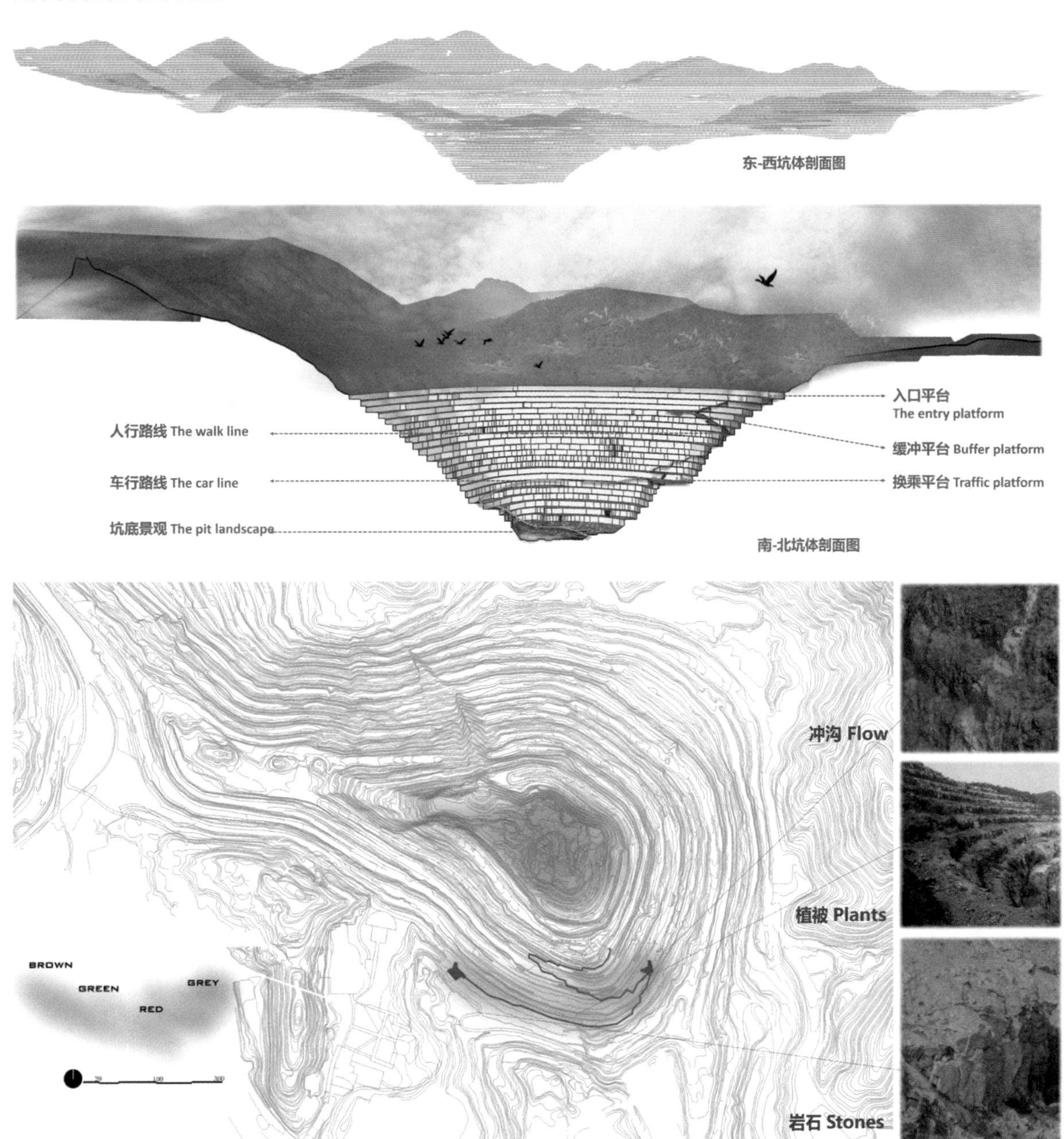

对立 / 连接

Contrasts and Connections

刘苡辰　樊宸　何茜
Liu Yichen/Fan Chen/He Qian

正如我们所知，自然给予我们丰富的宝藏，依赖这些资源我们得以生存。人类的力量是伟大的，运用智慧挖掘矿藏，让我们的生活质量得以飞速的提高。矿山那边的城市，在此消彼长中，迅速崛起。随着不断地开挖，城市与矿坑之间是一种此消彼长的关系。此时，矿山以自身带动了城市的发展。

从矿山到矿坑，无法回避的一个问题——人类在创造工业文明的同时，资源渐渐走入枯竭……人类的力量足以让大地面目全非。

其实，看到今天的矿坑，我们难免会感到惋惜。然而，重要的不是那些已经消失的，而是如何去更好地利用我们现在所拥有的。自然的力量，并不只有矿产一种。

矿坑，是否能以一种新的身份涅槃？带着这个想法，我们进行了大胆的尝试。

As we know, nature, to give us a wealth of treasures, depends on these resources we can survive. The power of man is great, and the use of wisdom to mine the mineral deposits, so that the quality of our lives can be improved at a rapid speed. Along with the unceasing excavation, between the city and the pit is a shift in the relationship.

At this time, the mine to drive the development of the city.In fact, today to see the pit, we will inevitably feel sorry. However, what is important is not what has disappeared, but how to make better use of what we have now. The power of nature is not only a mineral.

01 场地现状 Site Analysis

分层分析
Layers Analysis

场地现状
Site Context

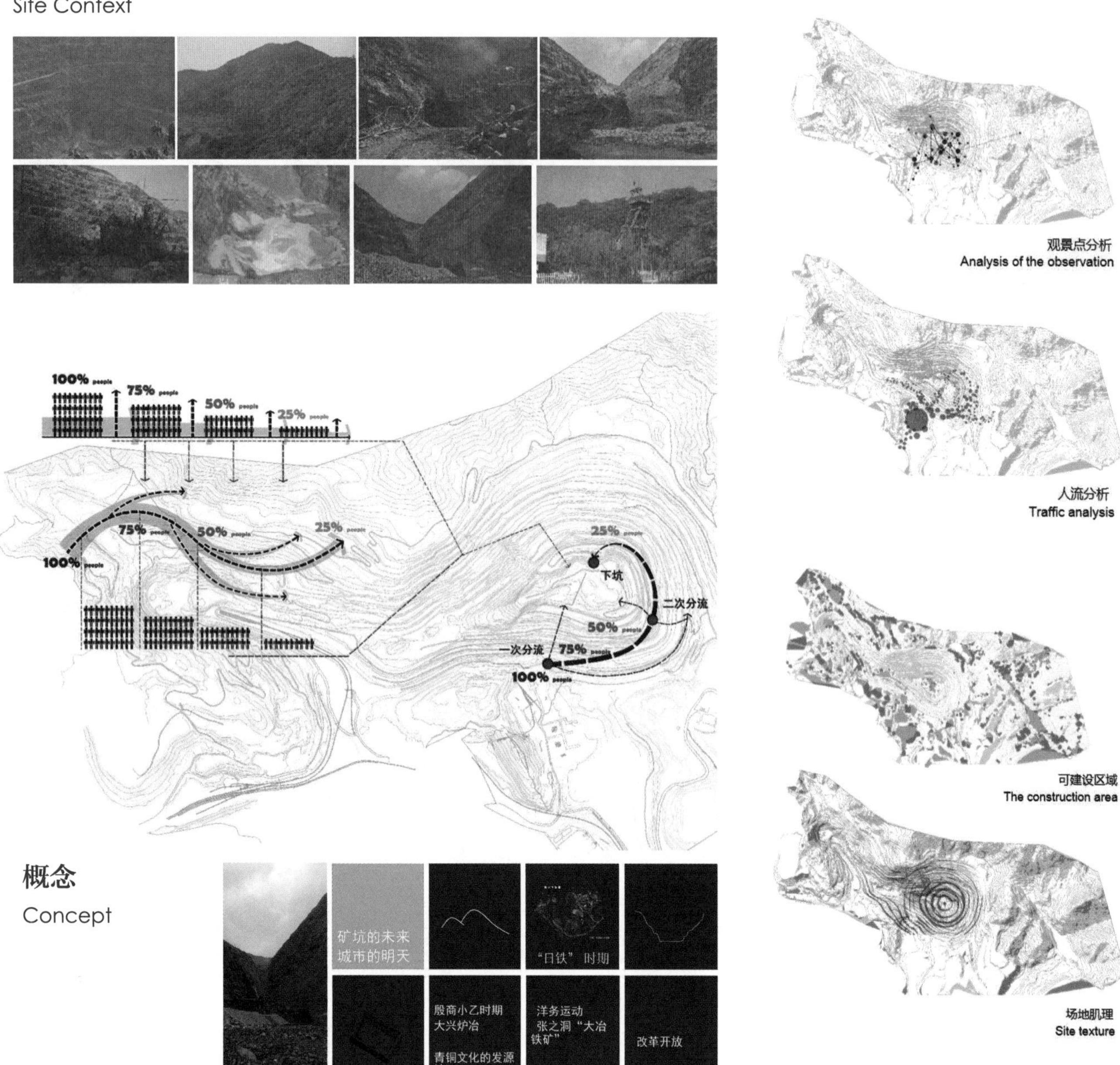

概念
Concept

黄石国家矿坑公园总体规划平面 Master Plan of Huangshi Mine Park

我们对矿坑的认识，经过这一个学期以来的讨论包括现场调查，我们一直在思考矿坑未来应该是怎样的身份，探讨人类、自然资源之间的关系。正如我们所知，自然给予我们丰富的宝藏，依赖这些资源我们得以生存。人类的力量是伟大的，运用智慧挖掘矿藏，让我们的生活质量得以飞速的提高。但是无法回避的一个问题，就是矿产资源始终是有限的。

03 详细设计 Detailed Design

游览节奏与路线
Tour Rhythm And Route

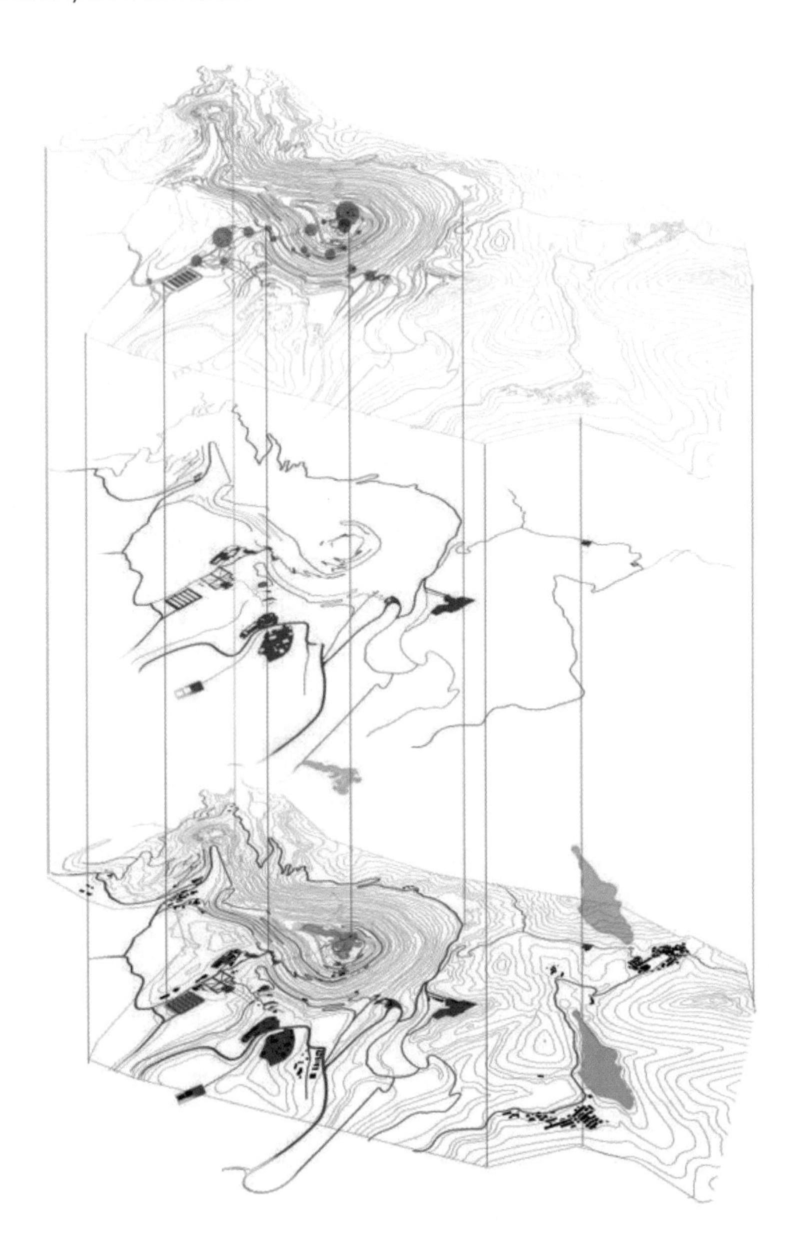

消蚀与生长
The Ablative And Growth

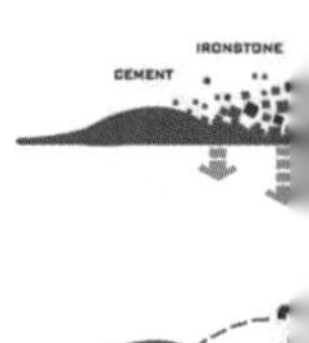

矿坑哺育了城市的生长。矿坑与城市的发展对比形成了绝佳的隐喻。矿产的不断消失为城市的发展提供了动能，这是一个此消彼长的过程。

Mine has nurtured the growth of the city. Mine is compared with the development of the city formed a perfect metaphor. Mine the disappearing provides kinetic energy for the development of the city, This is a reciprocal process.

快速路线
通过利用地形的特点和人群分流的计算最终得到快速路线。路线的起点前接入口，出口连接水坝。
FAST ROUTE
BY USING THE CALCULATION OF THE CHARACTERISTICS OF THE TERRAIN AND THE CROWD SHUNT EVENTUALLY GET FAST ROUTE . THE STARTING POINT OF THE ROUTE BEFORE ENTRANCE, OUTLET DAM

城市形态演变

Urban Morphology Evolution

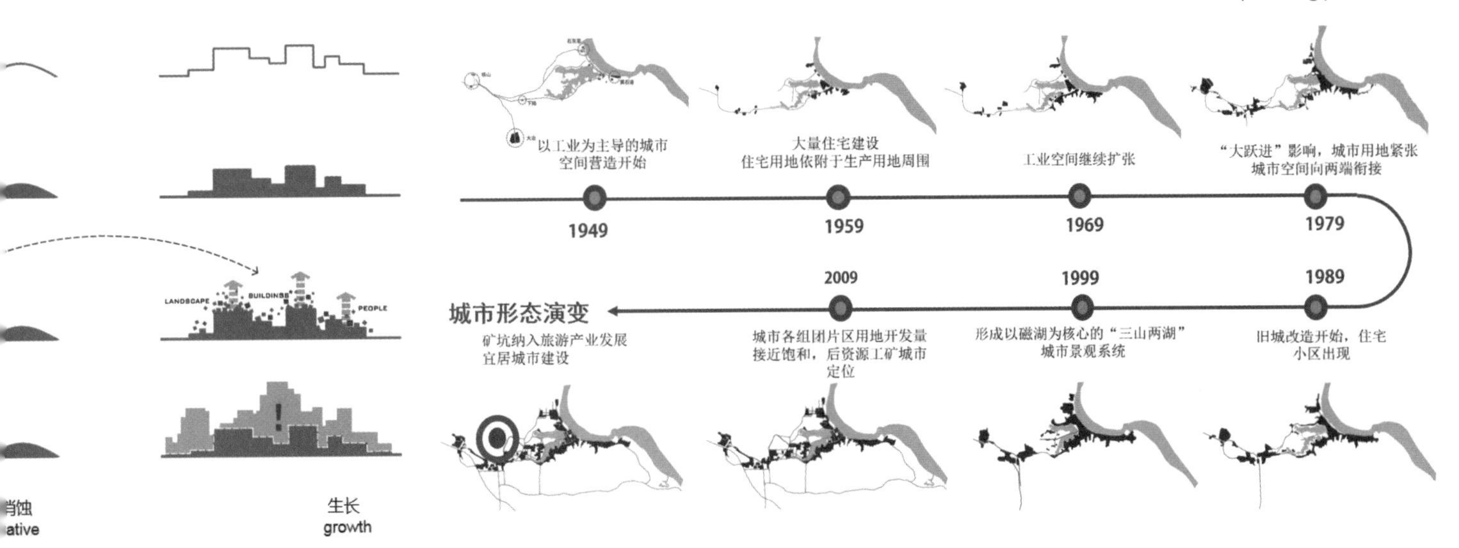

废弃场地景观设计

LPAM

李雪飞　王磊　刘永欢　张冬
Li Xuefei/Wang Lei/Liu Yonghua/Zhang Dong

调研路线依次为水泥厂房区、旧采矿区（现为遗留废弃平地）、新采矿区，沿路通过照片与手绘记录场地信息与感知。水泥厂建筑区带给人强烈的工业感和工业尺度，需要我们延续这种工业景观创造的不常有的感情以及可能包含的诸多功能；旧采矿区因为原有山体矿石被挖走，现状遗留的是废弃的平地，平时只有运料车经过，我们将试图寻求某种合适的功能以及交通规划同时更重要的是完成生态修复；新的采矿区保留了经典的矿坑形式，空间限定较为明确，适宜在生态修复基础上赋予更多的可供人活动的空间。

Research route is followed with the cement plant area, the old mining area (now abandoned flat), and the new mining area, along the road through the photos and hand drawn we records of information and perception. Cement factory building area show us a strong industrial and industrial scale feeling, we need to extend the industrial landscape for the unusual feelings which contains much functions; because of the original mountain ore was excavated, the legacy status of old mining areas is flat. We will try to seek a suitable function, the traffic planning and ecological restoration completed; the new mining district retained the classic form, the space qualified is clear, suitable for ecological restoration based on requirement human activity space.

01 场地现状 Site Analysis

场地现状

Site Context

根据现场调研情况，选取水泥生产区、运输过渡区、石料采集宕口区作为规划范围。

The range of the planning includes the fact-ories, the transition area, the entrance of the quarries, and the continuous quarries.

建筑区及过渡区规划分析

The Ablative Planning Of The Building Area And The Transition Area

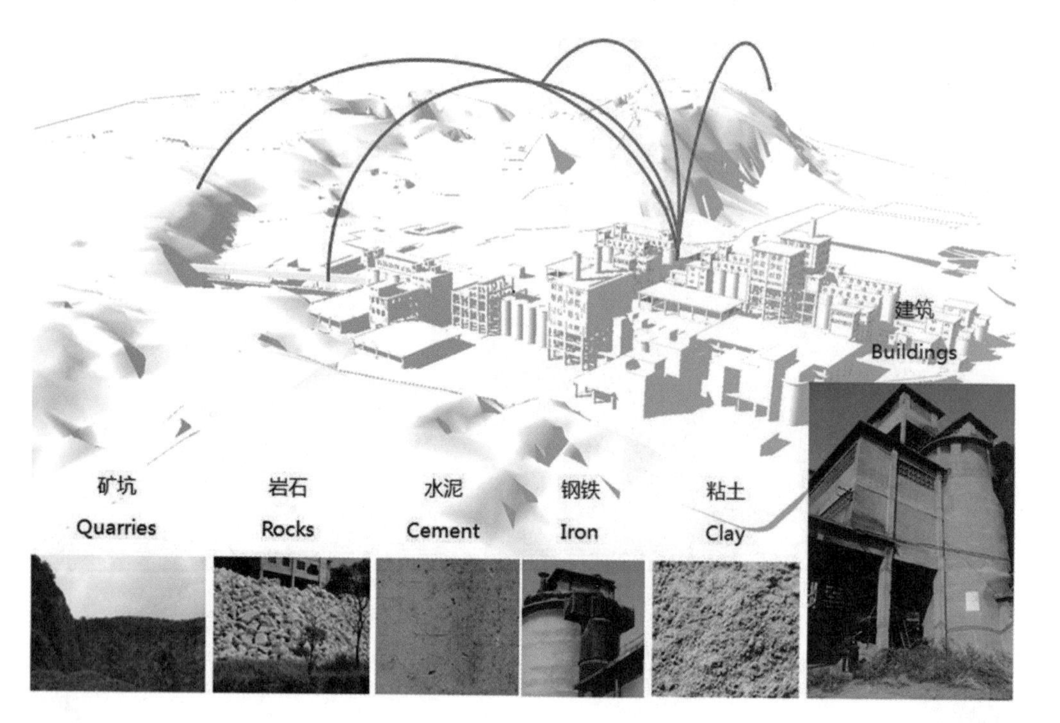

整旧如旧——保留工业元素，延续场地记忆。

Keep the old as being old, and keep the memory of the site.

新旧并置——新加元素符合场地原有精神，完成区域更新目标。

Coordinate the new and the old, keep the spirit and accomplish the transformation work.

筑巢引凤——预留可能的发展空间和潜力。

Reserve for all possibilities.

总体分析

General Analysis

建筑体内部人行串联不顺。

Difficult to linear the tour continuously inside the buildings.

建筑结构带有极强的工业特点，常规业态应用难度较大。

Hard to be used in normal way.

工业设备拆除难度较大。

Difficult to tear down all the equipments left inside the buildings.

结构较为简单的仓库建筑反倒具有较强的建筑利用可能。

More possibilities for the warehouse buildings.

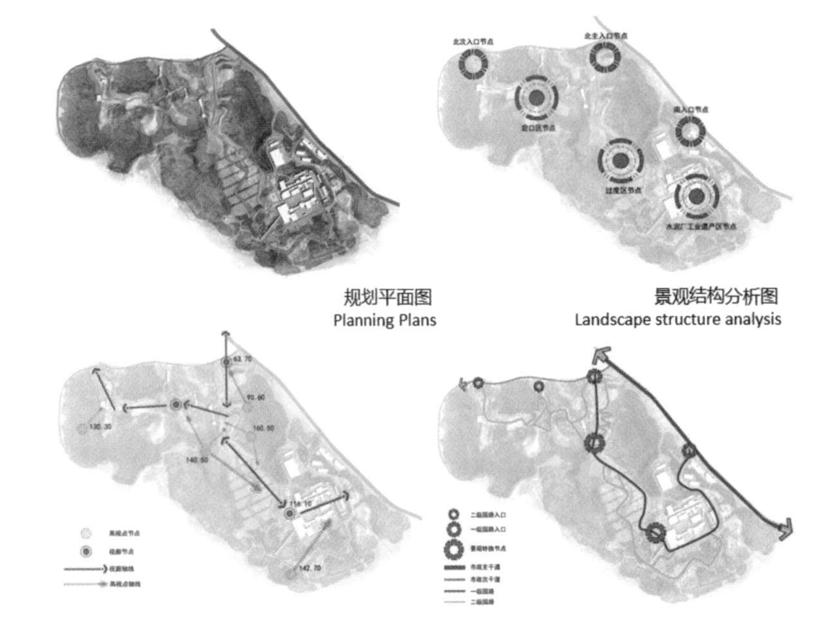

规划平面图
Planning Plans

景观结构分析图
Landscape structure analysis

景观视线分析图
Landscape sight analysis

交通流线分析图
Traffic flow line analysis

总平面

General Plan

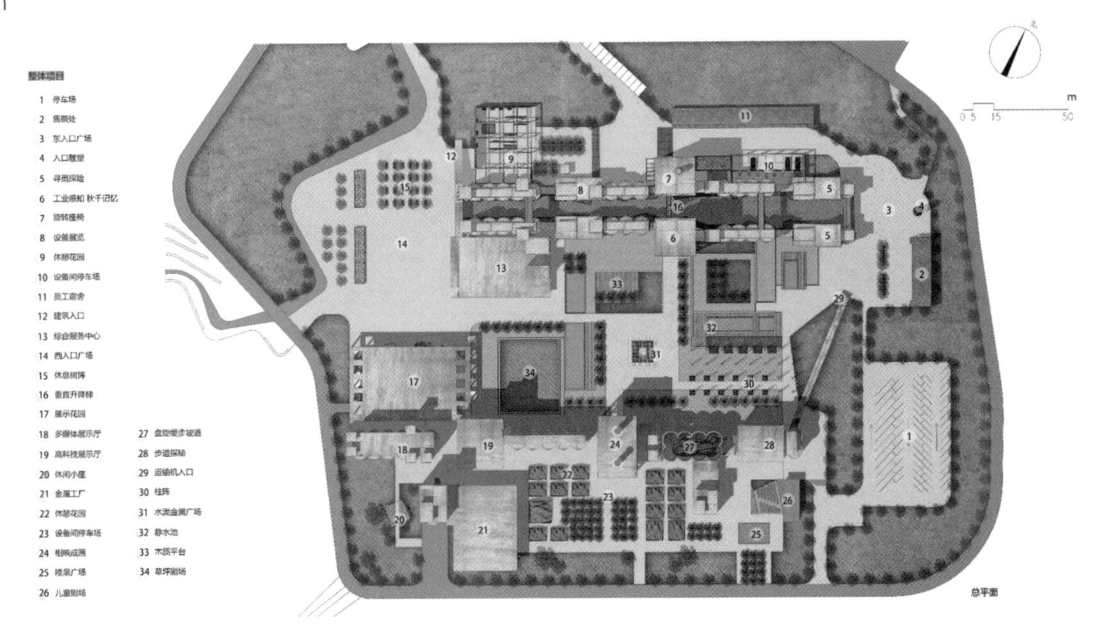

总平面

建筑南侧场地设计方案

Design Of The Southern Part Of The Buildings

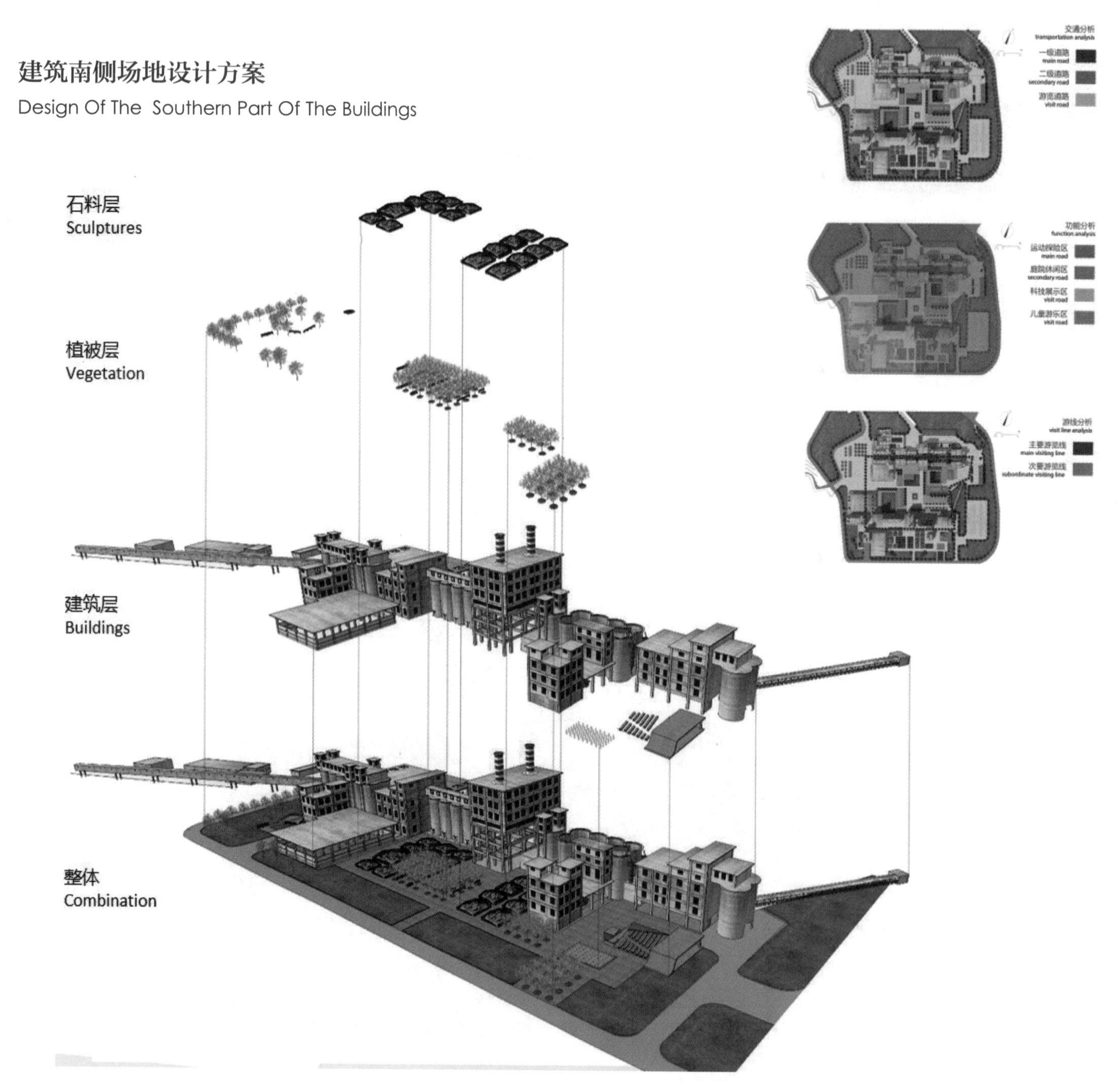

厂区中部改造

Rebuilding Of The Center Site

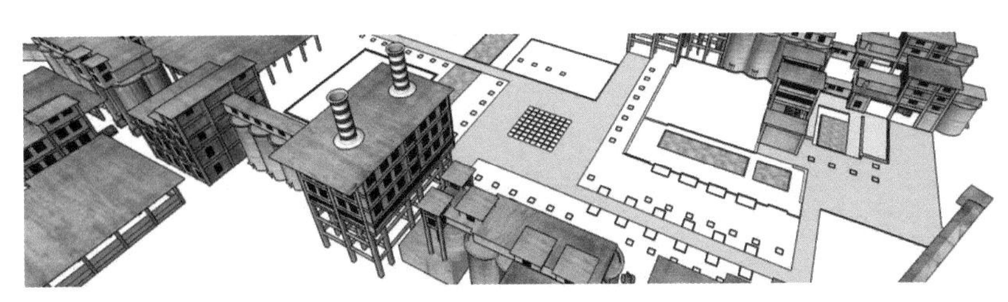

水池
pool

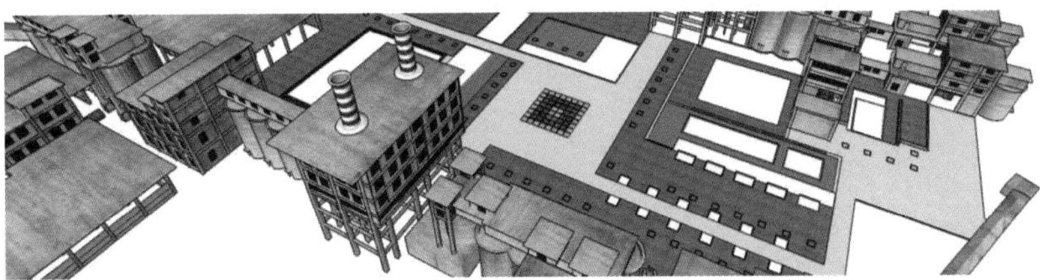

铺装
pavement

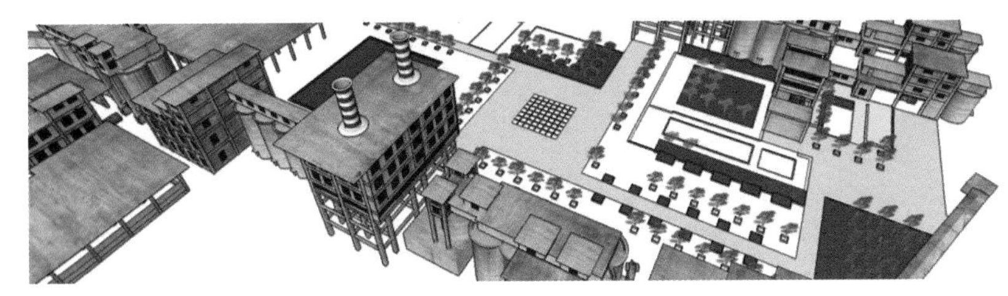

植物
plants

库房最外围一圈的柱子被设计成柱廊，且柱子被新的清水混凝土表皮包裹成圆柱状，柱廊内部用落地玻璃围合室内空间，按照游客服务中心的功能需要设计游客服务台、餐饮、纪念品等商品售卖等区域。

The columns in the outer range of the warehouse are designed into a colonnade, with the columns wrapped into a cylinder shape. Inside the warehouse, the full height glass are employed to border space. Tourists service counter, bar, souvenir shop are arranged in this building, according to the functional needs of a tourists service center.

游线设计

The Visiting Tour Design

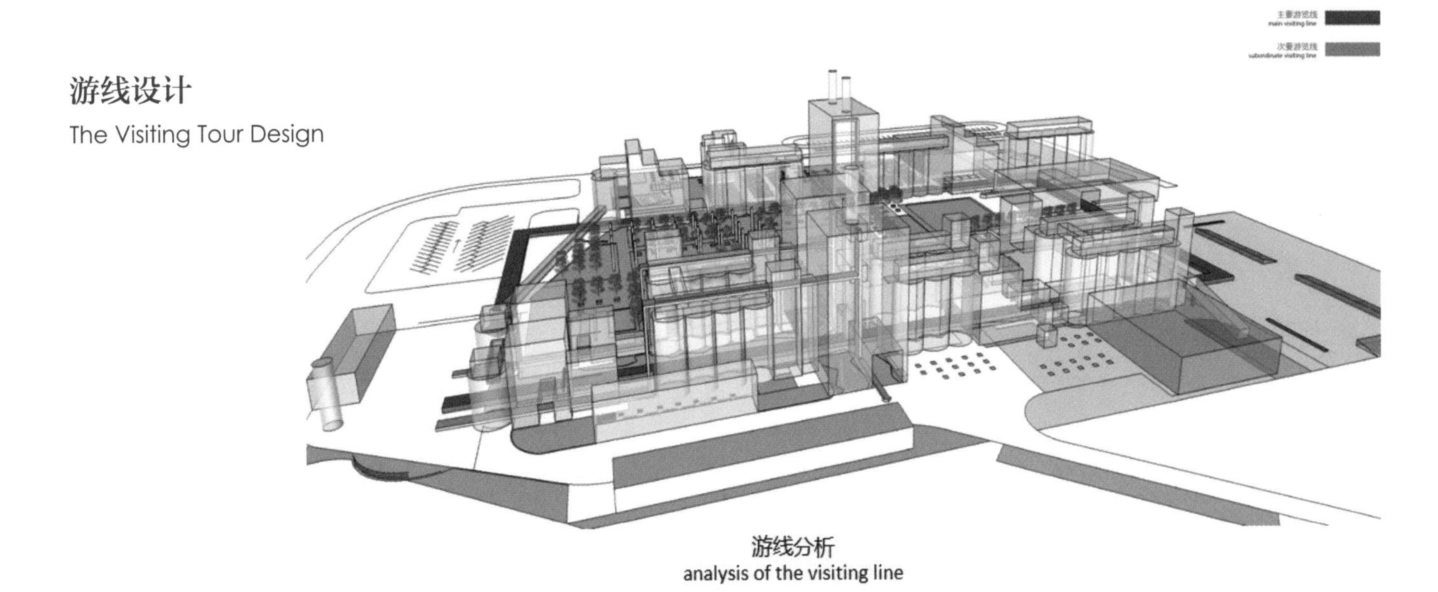

游线分析
analysis of the visiting line